Research Strategies in Technical Communication

Lynnette Porter
William Coggin

Research Strategies in Technical Communication

John Wiley & Sons, Inc.
New York • Chichester • Brisbane • Toronto • Singapore

Associate Publisher: Katherine Schowalter
Editor: Theresa Hudson
Managing Editor: Mark Hayden
Text Design & Composition: SunCliff Graphic Productions

This text is printed on acid-free paper.

Library of Congress Cataloging-in-Publication Data

ISBN: 0-471-11994-6

Printed in the United States of America

10 9 8 7 6 5 4 3 2 1

Contents

CHAPTER TWO Idea Generation as You Begin Your Research 19

CHAPTER THREE Planning Research: Documentation Plans 45

CHAPTER SIX Computerized Secondary Sources of Information 119

CHAPTER SEVEN Personal Experience and Knowledge 137

Preface

Technical communicators, as a group of professionals, work within the broad realm of technology, science, business, and communication. We may directly or indirectly be involved in educating, training, or preparing others to do their jobs better and to understand more about what we do. Regardless what area we work in or what types of job classifications we might have, we must have a variety of research skills to perform our daily jobs and to advance our profession and allied disciplines. With this book, we have attempted to provide a guide to research within *all* technical communication: from practical, day-to-day research techniques needed by technical writers to theory-based empirical research techniques needed by researchers and lab assistants. Although not every part or chapter may apply to your research interests and practices, at least some parts and chapters will provide you with strategies you can use in your career, whether you are a novice technical communicator or an experienced professional.

The book is divided into four parts:

Part I. Introduction to Research Strategies

Part II. Ways to Conduct Research

Part III. Ways to Evaluate Research

Part IV. Possible Publication of Your Results

In Part I, we emphasize strategies for getting ideas for your research, finding out what already is known about a subject, and recording the information on which you'll base your research methods. Documenting your ideas and projects is an important part of this process, so we describe a documentation plan as one way to illustrate your plan for completing a project and ensuring that all who work on that project know their responsibilities and deadlines.

Of course, ethics is an important part of any area of technical communication, and technical communicators should act ethically. As an overview of what is required ethically in our profession, a separate chapter about ethics provides you some guidelines to ethical behavior required of all researchers.

In Part II, we describe different approaches to qualitative and quantitative research. The day-to-day research methods needed by technical communicators who have to produce information for distribution to others are described. The empirical methodologies and associated terminology needed to conduct original research, to increase knowledge about a subject, and usually to test a hypothesis also are discussed. A valuable approach to locating secondary sources is computerized information retrieval. E-mail, the Internet, and the World Wide Web, for example, are becoming commonplace tools in technical communication, although not all technical communicators have as much access to computer technology as they would like. Thus, the strategies, illustrations, and guidelines provided in this chapter can form a foundation upon which you build additional computerized research skills.

In the chapters in Part II, we also discuss some methods of gaining information directly from people rather than from published sources. Informational interviews, for example, are a primary means of gathering information from subject matter experts, and one chapter is devoted to just that topic. However, other important strategies for technical writers, such as relying on personal experience and using prior knowledge as sources of information, are also included. Another way to gather new information is to write a letter of inquiry. Still other methods of conducting research include designing and administering questionnaires or surveys. Each of these methods is the focus of a chapter.

Of course, technical communication as a profession is made up of more than technical writers whose research is used primarily to create information that others will use very practically. Technical communicators who conduct research for its own sake are also an important group within the profession. These technical communicators may use their interviewing skills, as well as personal experience and knowledge, to gain new insight, but they may also need to develop an experiment to test their ideas. Or they may want to replicate someone else's work to test a research tool or method or to provide a basis for continuing another researcher's work. Therefore, a separate chapter deals with experimentation and provides an introduction to the scientific method of inquiry.

In Part III, we explain ways to evaluate the results of research. If you have gathered practical information that can be used in creating a document, for example, you may need to compare and evaluate the secondary sources that underpin your work or the primary sources that form the foundation for new material you'll share with others. Sometimes this information conflicts, or you receive a wealth of information from different sources, and you may be unsure of how much emphasis you should give to a particular perspective. Ways to evaluate sources are described in one chapter.

If your research provides you with raw data, you need to evaluate them, too. You may need to have a statistical expert help you determine which tests are best suited to evaluate your data, based on the type of research you conduct and the form of your data. In an introductory chapter to statistics, you can read definitions of the most common statistical tests and the terms used to describe your data.

Conducting usability tests is a research procedure that involves evaluating how usable a document, product, or process will be to the people who will use it. In a separate chapter about usability tests, you see an outline of the general steps involved in planning such a test. With this background, you should be able to develop an appropriate, specific usability test to help you learn about the effectiveness of the product being studied. This information can then guide you to conduct further, ongoing research or to make specific recommendations to improve the current product.

In Part IV, you read about ways to share what you have learned from your research. Some technical communicators may want to publish articles or books; others may make conference presentations. In this section, you read about ways to propose publishing your research by writing a prospectus and query letter. As you share your work, you might also need to prepare abstracts and bibliographies, or you might need to refer to others' abstracts or bibliographies to get ideas for future projects. Each of these topics is the subject of a chapter.

Research is a continuous process. It involves gathering information to complete specific projects, but it also involves remaining generally informed about the profession. It involves not only learning from others but also exploring your theories by experimenting with and writing and speaking about those theories. If you're working for a company, you may have support for this research, but if you're not, you may need to seek support. In Part IV, we describe some ways you might expand your research or receive additional support to conduct ongoing or new research projects.

Throughout most chapters you'll see references to online information. Although two chapters exclusively concern online information, computer technology is often used in some part of just about every research strategy that technical communicators use. Therefore, you will find additional information about computer technology as an integral part of several chapters. The examples, references, and descriptions of networks, the World Wide Web, mailing lists, and other computerized sources and forms of information reflect commonly used technologies available in early 1995, the time this book was sent to press. Because technology is changing so rapidly, newer information will probably be available to you as you read this book. The information in this book is designed, therefore, to help provide you a foundation for your online research strategies.

Each chapter includes descriptions and examples to help you learn more about a research method or strategy. You can practice the strategies or learn more about the concepts described in a chapter by completing the exercises listed at the end of each chapter. The end-of-chapter reading lists (For Further Reading) provide information about hardcopy sources, although some undoubtedly also can be retrieved electronically. Several other good sources about each topic can also be found, but we tried to include representative readings that can serve as a starting point for additional reading.

A list of some useful hardcopy sources about technical communication has been compiled from the For Further Reading sections into an appendix. Additional appendixes offer lists of sources to help technical communicators with their ongoing research.

This book is an attempt to emphasize the necessity of research within the technical communication profession. It is dedicated to all of us who conduct research in any way and in any discipline and to those who will continue to advance our profession through their ongoing research.

Acknowledgments

We thank the following people for their assistance in the preparation of the manuscript: Clifford Logan, Laura Mandzak, and Marte Salme from The University of Findlay and Harold Stonerock from Bowling Green State University's Computer Services.

PART ONE

Introduction to Research Strategies

If you're an experienced technical communicator, you're probably aware of the need for practical, day-to-day research that allows you to develop information that will be made available to the audiences who rely on it. As well, you understand the necessity of the ongoing research that helps increase knowledge about technical communication or one of the many technical, scientific, or communication subject areas that make up the profession. Therefore, if you're an experienced technical communicator, you may want to skim through the first few chapters to review what you know about research within technical communication. However, because ethics is a subject that transcends research methodologies, you may want to pay special attention to Chapter 4, Ethics. It provides a checklist for conducting research ethically and highlights the ethical procedures all technical communicators should follow.

If you're a new technical communicator or a new researcher, or just someone who's anticipating the need to conduct research, Part I provides you with important background information about the nature of research in technical communication, terminology that applies to all areas of research in our field, and examples of typical research scenarios. You, too, should find Chapter 4, Ethics, especially helpful as you begin to develop research strategies and work with others who conduct research.

Part I consists of these chapters:

- Chapter 1, Introduction to Research
- Chapter 2, Idea Generation as You Begin Your Research
- Chapter 3, Planning Research: Documentation Plans
- Chapter 4, Ethics

Chapters 1, 2, and 3 are most effective if you read them in order. They provide you with an overview of research in technical communication, ways to generate ideas for your research and to begin your research, and ways to develop a project plan and a method of documenting your work. Chapter 4 contains information that is useful throughout the research process. Although you should consider Chapter 4 an important part of the background information you need to know as you first develop your research strategies, you may want to refer to it periodically as you work with other chapters and conduct your research.

CHAPTER ONE

INTRODUCTION TO RESEARCH

If you're active in the profession of technical communication, you probably spend a large percentage of your time conducting research. In college or graduate school, much of your research probably involved going to the library to find sources for research papers, book reports, speeches, or presentations. During an internship, a co-op, or a first job in technical communication, you may have been exposed to fieldwork as part of your research. You may have made site visits or clinical observations. You may have conducted environmental tests, surveys, and interviews. You may have worked in a laboratory or a research center. All of these tasks are typical within our profession.

As you've probably discovered, research is a necessity in technical communication. It is an ongoing, important part of every technical communicator's professional and personal development, because the various sciences and technologies change rapidly, and each of us has to keep up with what's new and what's being studied in our technical or scientific specializations. When you are a thorough, efficient researcher, the research part of your job can be enjoyable as well as enlightening.

As you conduct research to complete projects, you may find yourself working alone: reading, writing, testing, or evaluating materials, processes, or products. At other times, you may be in constant contact with others at one or more stages of the process that takes an idea from conception to presentation. Regardless, you are engaged either directly or indirectly in a collaborative effort. Two common situations in which technical communicators conduct research are the following:

- They work within a company or serve as freelance writers or consultants for a company that produces technical information. The information may be in hard copy (paper-printed copy), such as manuals, reference books, feasibility studies, grant proposals, or other products of technical writing and editing. The information may be exclusively in soft copy, such as online documentation, computer-based training programs, interactive video "lectures," or CD-ROM bibliographies. Whether the resulting product is hard-copy or softcopy information, the technical writers and technical editors within these companies are responsible for working with subject matter experts (SMEs) within the company. These writers and editors produce the appropriate forms of technical or scientific information suitable for a given group of readers or users. As part of technical writers' research to produce the appropriate information, they may work with boilerplate from previous corporate information, conduct interviews with SMEs, use their experience and knowledge of the subject matter, and/or observe or experiment with the product about which they will write. This type of research is pragmatic and result oriented. It requires an immediate gathering of information and evaluation of the sources, so that the information can be used to create a product.
- They work as researchers for or with a university or college, research institute, or division of a company dealing with research and development (R&D). Their time is dedicated to research activities, such as developing hypotheses, designing research methods to test these hypotheses, conducting tests, gathering information from multiple sources, analyzing results, reporting findings, and studying the findings of others as the basis for ongoing research. This type of research may produce written products like journal articles, conference papers, and scholarly books, or it may lead to applications to advance knowledge in a discipline, improve a product, or solve a problem. However, the purpose of ongoing research is to add to the body of knowledge about a subject matter. This type of research is involved with theory more often than it is with application, and it may require years of information gathering and analyses before the results are shared with others.

Both situations are common in technical communication, and both are important to our profession and our international community of citizens. Both require researchers to have knowledge of at least one scientific or technical area or specialization, an eagerness to keep learning about that area and other subject areas, good communication skills, and an analytical mind. Research also requires continuous learning about constantly changing sources of information storage and retrieval, for example, networked databases and other electronic media. To be an effective researcher in technical communication, you must be open minded at times, analytical and critical at others, always methodical in your approach to the research, and continuously interested in the research process itself. You should also be curious, because the best researchers are curious people who want to learn more.

Technical communicators involved with research in different scientific and technical areas and working in different research situations may have very different job titles. They may be technical writers working within a company's documentation department, technical editors responsible for editing all information produced by a company, freelance writers gathering information from interviews, or consultants relying on their expertise and experience they've gained by working for several clients. Technical communicators whose primary job is conducting research may be called lab technicians, product-development specialists, or interviewers. They may be authors of books, articles, or bibliographies. They may be inventors of new processes or products. To create a standard title to encompass this variety of interests and responsibilities, the terms *researcher* and *technical communicator* will be used throughout this book to indicate a professional technical communicator doing any of a variety of research activities.

Throughout this book, you'll work with different research methods and strategies to help you work well in one, if not both, types of research situations found in our profession. Although you may now or someday specialize in one type of research, you should have a variety of research skills that enable you both to understand and apply research methods to different situations.

Qualitative and Quantitative Research

Two general categories of research in which you may be involved are defined by the types of results appropriate for a particular discipline or situation within a discipline. Sometimes, for example, research is conducted to result in countable data; at other times, the objective is to seek impressions or descriptions, for example.

Quantitative research involves methods that produce countable results. Enumerated or counted information then can be ranked to determine highs and lows, averages, and the ranking of one item when compared numerically with another. For example, if you developed a questionnaire that consisted of several multiple choice questions, you could count the number of people who answered a question and those who did not answer a question. You could also count the number of people who selected each possible response. Then you could compare the numbers of responses, as well as the number of people who answered a question with the number of those who did not answer it or those who provided an answer different than the choices provided for that question.

Qualitative research involves more descriptive methods and the categorization of information. Categorizing events, such as job tasks, the ways people respond to a question asked by an interviewer, and the words people use to describe their feelings or behaviors, is the basis of qualitative research. The researcher's observations of an event are recorded and labeled so that they fall into clearly established categories. For example, if you wanted to learn how a group of document users work with a document that's just been created, you could observe the ways they use the document to complete a task. You could have them follow the instructions for entering data into a computer program, for instance. Then you could observe each person as he or she completed the task and record each person's comments about the process. The descriptions of each person's activity could then be labeled and categorized. The types of comments could also be identified, labeled, and categorized. Then you could compare the categories, as well as learn how well people could follow the instructions by focusing on particular comments or observations.

Types of Research by Discipline

Technical communication involves many disciplines. Some sciences may require technical communicators to conduct empirical research. Laboratory and field research in biology, chemistry, physics, and their interrelated specializations often deal with hypothesis testing, information gathering, and analyses of results. Studies in the social sciences, such as psychology and sociology, may require technical communicators to make clinical observations or design questionnaires as part of empirical research. Case studies, observations, and individual and group interviews may be some research activities in the humanities, through studies in art, communication, literature, history, music, and rhetoric, for example. Research in technological areas, such as robotics, engineering, and construction, may involve theory-based research that leads to practical applications.

Depending upon the purpose of the research and the type of information you want to gather, you use qualitative or quantitative research methods within any of these disciplines. The type of research method you ultimately use reflects careful consideration of the project's purpose, the type of information you need to gather, and applications of the results. That's why you should at least understand the basic vocabulary associated with and the principles of using both methods.

Within each specialization in the profession, you usually find specialized guidelines or traditional methods that can help you hone your research skills when you conduct research within that specialization. For example, you may discover that the researchers working at a particular institute and conducting research within one specialization have developed specific research procedures that have become standard. When you conduct research for that institute and within that discipline, you might use only quantitative research methods, and you might follow strict guidelines for developing your methodology. When you're developing a research plan to test your hypothesis, you may have a great deal of latitude in determining which type of research, qualitative or quantitative, is appropriate for your project.

Research Skills Needed in Technical Communication

If you've primarily worked with just a few types of documents, you clearly understand the research strategies needed to gather the appropriate information to create those types of documents. If you've worked with R&D in one company, you're familiar with the research methods, usability testing, and marketing research that are successfully used within that company. As a student, or as a practicing professional, you have experience with library research and perhaps some empirical research methods.

Each experience in research helps develop skills that can be applied in different situations, but skill at practicing research methodology does not alone make one an excellent researcher. In general, the skills necessary for technical communicators to conduct research, whether for the immediate production of technical or scientific information or the longer range testing of hypotheses, and some characteristics that help make the application of those research skills most productive include these:

- Curiosity: to learn more, to discover why—or why not, to follow up on questions, to ask new questions, to wonder.
- Interest in detail: to document everything, to work methodically, to check facts, and to discern among facts, theories, hypotheses, and opinions.
- Interest in fact: to search for the truth, to analyze, to check and double check, to test, to validate, and to replicate.
- Ability to see trends: to envision the future, to anticipate needs, and to take the next step.
- Awareness of audience: to understand needs and expectations, to empathize, to assist, and to plan what should be done next.
- Critical thinking: to look as objectively as possible, to verify, to support, to analyze, to criticize, to evaluate, and to think logically.
- Innovative thinking: to look as creatively as possible, to try something new, to try something different, and to make connections among ideas.
- High ethical standards: to expect logical, honest, and collaborative work from yourself and others; to credit others' work; and to conduct safe and appropriate research.

To be a researcher is to be a scientist, a documentor, an inventor, a critic, and a visionary. It requires the ability to be left- and right-brained. Most important, it requires the ability to develop an effective, immediate strategy for conducting the research with an ongoing appreciation of new information and a long-term interest in the importance of gathering information and putting it to appropriate uses.

Primary, Secondary, and Tertiary Sources of Information

As you begin to conduct research about various topics, you'll discover that the process will often pose two distinct problems. For some topics, finding information is like searching in a bottomless chasm; there appears to be a dearth. With other topics, however, finding information is like climbing a mountain. Wherever you look there's something to be studied and learned.

The latter case frequently presents more problems for you. Where there is much information, you need to know where to begin so that you can find the information you really want and can manage the information you find. If you want to climb the mountain, you don't need to start out at the bottom and slowly walk your way around and around, gradually climbing toward the top and seeing every pebble along the way. You pick a place and begin to climb, taking a safe, direct route that lets you see what you need but stay away from the dead ends and sharp cliffs. In climbing the mountain, you may take the most direct route, then backtrack indirectly if the climbing gets too tough. In information gathering, you do the same. You plan the route you want to take, so you're less likely to face a difficult obstacle or get lost along the way, but you always stay prepared to change the route.

In conducting research, you often find that once information is created, it appears to be self-perpetuating, and related information about a subject grows so rapidly that the subject becomes, in effect, a mountain. For example, if you look for sources about the Internet that were written five years ago, you probably find very few. But today, if you look up *Internet*, you find more sources than you could possibly read. The first sources of information about this subject were well received and responded to a growing interest in the topic. More information was created, which led to more interest and more resources being created. A mountain of information practically exploded onto the scene, and now readers are threatened with an avalanche of sources about this topic. Although it's good to have so much information available to you when you need to research the Internet, you might feel overwhelmed if you lack a good plan for dealing with the number of sources and the varieties of available information.

When this situation happens, you need to remember that as information grows from its source, the information becomes less and less usable, and some sources are not as reliable as others. For that reason, researchers have divided the kinds of information available on any subject into three distinct categories, according to their distance from the original source of the information: primary, secondary, and tertiary sources.

A *primary source* is information developed by the researchers themselves. For example, if you're a researcher who creates a questionnaire, conducts an interview, or observes an event, you're using a primary source. You developed the information based on your knowledge, experience, and actions. In technical communication, this source of information is the most important, because much of what needs to be written or transmitted to our audiences is state-of-the-art information. It must be created specifically for a document, presentation, or product designed for a target audience.

A *secondary source* is information created, printed, and/or published by someone else. Researchers who replicate someone else's experiment, cite a journal article written by someone else, or write up someone else's notes, for example, are using secondary sources.

Secondary sources are often called library sources, because the published form of the information is found in a library. If you read a book, print information from an online bibliography, locate a city on a map, or view a videotape, for example, and you didn't write the information, create the graphic, or make the tape, you're using a secondary source. Library sources can be found in private libraries, in addition to libraries operated by a university, college, city, or county. Companies, chambers of commerce, museums, and professionals' offices often house technical or scientific information. Companies have documents they've previously published, either for in-house use or public distribution. The chamber of commerce typically has brochures, booklets, maps, and other information about a community and its businesses. Special collections may be found in museums and galleries; and offices of professionals, such as dentists and doctors, often contain a wealth of information about a specific profession or subject.

You've probably developed a library at home of textbooks, reference materials, handouts, disks, brochures, and other information accumulated through courses you've taken, conferences you've attended, and hobbies you've developed through the years. These secondary sources can provide valuable background information, but you should know where to locate these sources in a variety of different libraries.

Secondary sources can help you learn more about a subject, in case you're researching information outside your area of expertise. They can help you learn how much has been written about a subject, and by whom. They can provide you a history of a topic and critiques, commentaries, reviews, and analyses of others' work. They may have additional value as an authoritative source considered as a standard within a discipline. They may provide a valuable foundation upon which later work was based. Depending upon your research area, secondary sources may be important to your understanding of a subject, its history, and its significance, but you should carefully evaluate whether dated information has current applicability.

A *tertiary source* is information someone else created based on yet another's or others' information. For example, researchers who read a critique of a review of someone else's journal article are using a tertiary source. If you read a summary of the comments a group of users made about a manual someone other than you wrote, you're using a tertiary source. In technical communication, tertiary sources are less important than secondary and far less important than primary sources. The farther the researcher is from an original source of information, the greater the possibility the information will need to be updated or supported through other sources. Tertiary sources may provide useful background information that help the researcher gather some general information about a subject, but they most likely are not specific or recent enough to be useful in a current research project.

Table 1.1 lists some primary and secondary sources useful in technical communication research.

Research Methodologies

Research methodologies involve many activities using primary, secondary, and tertiary sources. Common research methods in technical communication include the following:

- library research
- personal experience and knowledge
- interviews
- letters of inquiry
- questionnaires or surveys
- experimentation

Part of your job as a researcher is to identify which methods will be most effective for your research. As well, you may need to use several methods, not just one, to gather all the sources of information you need.

Each of these methods will be described in detail in separate chapters. As a researcher, you may favor or work in an industry or a business that requires the use of some methods more than others. But you should develop an understanding of all these methods and a sense of the importance of careful, well-planned research strategies and methods.

Table 1.1 Some Primary and Secondary Sources of Information

Primary (conducted or created by you)	*Secondary (published or produced by someone else)*
interviews	books
questionnaires	journals
surveys	magazines
letters of inquiry	newspapers
experiments	newsletters
observations	brochures
personal experience	booklets
personal knowledge	government documents
	manuals
	reports
	instructions
	conference papers
	product information
	maps
	letters
	notes
	advertisements
	scripts
	speeches
	videotaped information
	audiotaped information
	filmed information
	computer disk information
	software
	photographs
	slides
	holograms
	games
	simulations
	online information
	online documentation

Library Research

Library research provides you with a wealth of secondary and perhaps some useful tertiary sources from which you can later develop primary sources of information. Library research helps you locate background information that may help you learn about a scientific or technical subject or to see how much information has been published about that subject within a certain time period. Trends in research or public interest in a particular topic can be deduced through library research.

In a library, you may work with computerized databases, reference works, government documents, journals, books, maps, records, letters, audiotapes, videodiscs, photographs, and so on. Knowing how to design research strategies to locate information in, and work with, online databases, softcopy information, and printed documents is a key to getting the greatest value from these sources.

Although the library may be in many different locations and include businesses', organizations', institutions', and personal collections of materials, getting to the library may be easier than ever. With electronic access to libraries and the ability to locate and download or print information locally, regionally, nationally, and internationally, the definition of *library* is changing. Much library research can now be accomplished from almost any site that has a computer and access to a network.

Personal Experience and Knowledge

Personal experience and knowledge are valuable primary sources of information, because no one else is like you. As a technical communicator, you have a unique perspective on the subjects you research, because no one else has accumulated the knowledge, life experiences, and work experiences in just the way that you have. In the profession of technical communication, you're expected to know a great deal about at least one technical or scientific subject area and to have gained and always be gaining experience in your part of the profession.

Nevertheless, because you've shared common experiences with others or probably have had life experiences that many others also have had or will have, you can use your experience and knowledge to help others. Your experiences and knowledge are useful when it comes time to apply what you already know to what you need to find out. Your experience working with a similar product or in a similar situation helps you when you have to write about a new product or situation. Your knowledge of theories, hypotheses, laws, and assumptions are part of the body of knowledge shared by experts in your specialization. This knowledge can help you design research to expand the knowledge base or to apply what has been learned to what needs to be done next. Your knowledge of what it's like first to sit down in front of a computer and learn to operate a new application, or what you wanted to know when you first saw a new model car, can help you provide this information for others who will be going through similar situations.

Interviews

Sometimes you may need to go straight to the people who have firsthand knowledge about a subject. At other times, you might just need a different perspective from someone who has a different idea about a subject or whose experience provides a different outlook than you've been able to learn about from other sources. At these times you turn to an SME. Usually you conduct interviews with SMEs to use their experience and knowledge, but you may talk with them to gather support for your research ideas or opinions about a subject. The information you gather is a primary source, because you create the questions, conduct the interview, take the notes or otherwise record the information, and use the information in your project.

Within many companies or for many freelance technical communicators, informational interviews with SMEs is the most common means of gathering information. Although face-to-face interviews, especially with in-house SMEs in a company, are most common in technical communication, conference calls, telephone interviews, and teleconferences are other ways to interview these experts.

Letters of Inquiry

Researchers who can't talk with SMEs in person or electronically often use letters of inquiry to gather information. This formal means of gathering information is useful when you need information from publishers, producers, or SMEs with a national or an international reputation. If you need a transcript of a television program, you write a letter of inquiry to request the transcript. If you have a question about an SME's comment in a recent journal, you write a letter of inquiry to clarify the information. If you want further research data indicated in an article, you write a letter of inquiry to the researchers to request a copy of the data. Because the recipients of these letters may take awhile to respond, letters of inquiry are useful when you have plenty of time in which to gather the information.

Less formal letters may take the form of e-mail messages, a popular—and faster—alternative to traditional mail. You still request information from SMEs, but you do it electronically and probably more often. The ongoing dialogue between you and SMEs can be a valuable source, kind of a cross between a personal interview and a formal letter. As well, if you request a document through e-mail, you may be able to print or use an electronic version of the document sent across the network.

Questionnaires and Surveys

Questionnaires and surveys are an important method of gathering information when the research requires responses from many people, perhaps with certain qualities, features, or experiences in common, or perhaps residing in a common geographic area. Questionnaires are longer and more complex, whereas a survey may be shorter and more informal, although the names are frequently used interchangeably.

You must carefully design questionnaires and surveys so that respondents can easily provide the information and you get the kind of information you need. This method of information gathering is useful to get ideas and opinions about a product or subject, conduct empirical research, or develop marketing strategies, for example—all which are important in different disciplines within the technical communication profession.

Experimentation

Empirical research involves the scientific method, a process of developing testable hypotheses, designing research to test the hypotheses, conducting the research, analyzing the results, and evaluating the research and the hypotheses. This approach is required for designing an effective experiment. Experimentation is especially useful in gathering information that can expand experts' knowledge about a subject.

Observation may be a crucial task within an experiment, especially within technical communication. You may set up an observation within a company, for example, so you can learn how employees typically do a job. You may want to observe real users' reactions to and use of information being usability tested. Or you may want to observe a mechanical process, such as the way a piece of equipment operates, before you write descriptions of the equipment or training materials to help people operate it. These situations may require you to observe from a distance or directly next to the person or equipment performing a task. Of course, as with all research, you must make observations in a manner that protects those being observed and provides you with accurate information.

An experiment, which may include or exclude observation, must be designed carefully and methodically so that the information gathered will be useful. In addition, the methods involved in experimentation may involve replicating the original work to see if the same results are obtained. Variations of an experiment may involve manipulating a variable to see what the results will be. Technical communicators who work in R&D departments or in research institutions, either within private corporations or academia, often conduct ongoing research and create many experiments to further our knowledge. Scholarly research is an important part of technical communication, and experimentation is an important part of scholarly research.

Summary

Research may be pragmatic or educational. It may be a daily task that's part of your communication process, or it may be a lifelong search for new information. Anywhere along the continuum of interest and necessity for research, you will probably develop a research specialty. You'll become familiar with the research practices common in your part of the profession. However, in addition, you should be familiar with the variety of research methods and interests included within the profession of technical communication.

Because technical communication is such a broad-based profession, it involves science, social science, humanities, and technology. It requires an ongoing interest in and facility with research, both in theory and practice. It demands an honest, ethical approach to conducting research. Throughout this book you'll read about very different research methods and emphases. It's up to you to develop a general understanding of research strategies and methods and a specific understanding of the types of research activities needed in your field as you develop your specific research interests.

Research Activities

1. List your research interests and possible topics for research projects in technical communication. Determine which project is possible to complete within your time limitations and geographic restrictions.

Create a database of research projects and ideas for future investigation.

2. Find a secondary source that includes or describes each of the following:

   - a questionnaire
   - a lab experiment
   - an abstract of a research report
   - a journal article describing a research project

   What do you learn about the research methodology by looking at this source? How can you determine the purpose of the research? What can you learn about the technical or scientific subject area being described?

3. Begin an informal online bibliography of secondary sources that describe your technical or scientific area of expertise. Add at least five sources to your bibliography each week.

4. Talk with other researchers within your company, institution, or professional association. How many different disciplines are represented by your contacts? What types of research do these people conduct? You may want to create a database of contacts you're making with other researchers, who can provide you additional insight into their work and the skills needed to conduct research within their discipline.

For Further Reading

The following secondary sources can provide you more information about the need for research in technical communication and current research interests in the profession. You should also look for additional information about research in your technical or scientific specialization.

Hernon, Peter. 1989. "Government Information: A Field in Need of Research and Analytical Study." *United States Government Information Policies*, pp. 10-13. Charles R. McClure, Peter Hernon, and Harold C. Relyea, eds. Norwood, NJ: Ablex.

MacNealy, Mary Sue. 1990. "Moving Toward Maturity: Research in Technical Communication." *IEEE Transactions on Professional Communication* 33: 197-204.

___. 1992. "Research in Technical Communication: A View of the Past and a Challenge for the Future." *Technical Communication* (November): 533-51.

Pinelli, Thomas E., and Rebecca O. Barclay. 1992. "Research in Technical Communication: Perspectives and Thoughts on the Process." *Technical Communication* (November): 526-32.

Smith, Frank. 1988. "The Importance of Research in Technical Communication." *Technical Communication* (May): 4-5.

CHAPTER TWO

Idea Generation as You Begin Your Research

In technical communication, the two most common research situations involve gathering information to fulfill the requirements of an assignment and conducting research that you design to learn more about a subject. You might be told, "Write an instruction manual for the software we're developing," "Write a proposal to request funds for new equipment," "Write a progress report for our client," or "Write an article about virtual reality and how it's being used in physical therapy programs." Or you might have told yourself something like "I want to write an article about superconductors," "I wonder if I can improve the engine's performance by changing this setting," or "I'd like to submit a grant proposal to study the use of teleconferencing to teach college courses."

Each situation has a different impetus, and each requires you to complete a set of research tasks. In the first situation, for example, maybe the assignment itself identifies who your readers are, why they need or want the information, and where they'll be using the information. You may have written similar documents for your audience, or you may be able to get this information from your client or company. Your research situation begins with refining the general information you already have about the audience, subject, and purpose—those factors that control content and presentation.

Your research is practical, because someone—usually a supervisor in a company or a client if you're a freelance writer—needs information, and you, as the technical communicator, are expected to meet that need. The person who gave you the writing task may suggest sources for your research or the direction your research should take, or you may know some good sources of information to help you get started.

The second situation is much different. In this situation, you begin with content—your idea for a project—and then analyze the audience and their needs. You have an idea about what you want to write, but you have to conduct research to make sure your information will be useful to an audience that you have yet to define. And you may be led into additional research to make sure that you can offer the right type and amount of information about the subject you've chosen.

You may have been given a subject about which to write a document. You may want to create a written product, such as an article or a proposal, based on your idea. Or, you may want to develop an idea for empirical research that, you hope, will advance the body of knowledge about a subject. In any of these situations, for example, you need to generate ideas, refine those ideas, and then develop a sound research plan.

In each situation, you have to generate ideas and keep an open mind about the project you'll eventually undertake. You can't just jump into research. You have to let your ideas develop, and you have to evaluate them carefully. Then you can better plan the tasks you'll need to complete as you begin the project.

Idea generation is an ongoing process, too. Of course, when you need to develop a good idea for a current project, it's good to know how to generate several ideas, which can be evaluated and refined. But idea generation in research is necessary even when you don't have a current writing assignment or you don't need to conduct research for a project you want to develop on your own. Good researchers are always open to new ideas, developing their thoughts and dreams, and keeping up with what's going on in their discipline.

Like many technical communicators, you may have specific research interests, which may or may not have anything to do with your daily job. For example, although you may be a research assistant in a laboratory, you also want to write a book about a very different subject than you deal with every day, or you have developed your own hypothesis you want to test. Some careers involve research in many different areas, and you're expected to have your own research interests, even if your daily tasks may not directly involve research. University professors, for example, may have the daily tasks of preparing materials for classes, teaching classes, serving on committees, and writing administrative reports, but they are also expected to have their own research interests. They may serve as consultants and research what others want them to research, or they may develop an experimental design on their own. As you work in technical communication, chances are good that, either by necessity or by interest, you'll be working in several different research situations, but you may find yourself in one situation more than another.

This chapter is divided into three sections. First, you'll read about idea generation when you are given a writing or research task. Next, you'll read about idea generation when you decide you want to create your own research design. Finally, you'll read more about computer-based idea-generation techniques, which can be useful no matter what type of research project you'll eventually complete.

Writing and Researching in a Company Environment

If you're a technical communicator in a company, you may be assigned most of your research and writing objectives, or they may be part of a continuing job function, for example, producing a periodic newsletter. Even if you're a freelance or consulting technical communicator, your client, whether an individual or a company, provides you with the projects you're expected to complete. The topic for your written product has been decided for you, as perhaps have many of the standards, policies, and procedures you'll follow, but you have to do the research needed for writing the product.

Producing a document for a company or other client is seldom an unconditional assignment. The research you complete and the process you use to complete it are directed significantly by time, personnel, budget, and facilities and equipment. Your research is also frequently affected by the stage of development if you're researching and writing about something that is still being developed and tested. In these kinds of situations, it's imperative for you to know immediately all information that affects your research. The following are examples of that kind of information:

- the starting date for the project
- the ending date for production or distribution of the product
- the time you can devote to the project

- the sources that are immediately available
- company policies for online and printed information, tone, style, and format
- the audience, subject, purpose, and scope of the information to be created

Technical communicators usually try to get this information during an initial meeting or a series of meetings with the team that will produce the product. They may already know some pieces of information because they've completed similar projects for the same company. They may be given the results of usability tests completed by real users of an earlier version of the product, which may be a document, or of similar products. They also know that readers or users want the information, but they have to determine how to design the information so that it will be accurate, timely, and usable.

Having this information saves technical communicators a great deal of time in conducting market research to make sure there's a need for the product or to determine what users think of the current product. Information provided by the company or client also pinpoints the audience and the scope of the information—two areas that someone had to research so that the information in final form will be useful and usable. If specifications, or specs, for the written product are provided, technical communicators don't have to locate and read company policies and similar documents created by that company in order to learn the specifications that will be required for the current project. This background research is simply provided, but its importance shouldn't be minimized.

Some companies also provide a list of subject matter experts (SMEs) who have been asked to work with technical communicators on a project. These people are probably in-house employees who often provide information about the products they're designing or developing. Technical communicators who have worked for the company on other projects usually have their own list of contacts who have been helpful in the past. This process saves a great deal of time in determining who will be good SMEs and then getting permission to work with those people.

Although a good amount of research may have already been completed and presented to you as a technical communicator working for a company, you still have other research to do. At this point, you may ask yourself, "Where do I start?" Although the project may be outlined thoroughly, you are expected to fill in the outline. Conducting research—

often in a very short time frame—is critical to the production, and, more important, to the success of the information. So, where do you start?

1. Start with what you already know.
2. Review what you've recently learned.

Start with What You Know

Starting with what you already know happens almost automatically when you've worked as a technical communicator for awhile. Technical communicators quickly become familiar with writing different types of information in different formats—some online, some in hard copy, some in multimedia—for different audiences—some for different employers, some for their own use and reflecting their interests, some for scholarly publication—and about different subjects. What you already know can be a wealth of information about what the product should look like, how long it will be, how much information will be new, how much information will come from boilerplate, and so on.

But it can also include familiarity with sources of information that have been useful in the past, when you had a similar project. You may know of SMEs within the company who can provide you the information you need, or you may know that you can experiment with the equipment or new product about which you'll write. You may be able to observe a job being done so that you can write better training or safety procedures. Because you've done similar types of technical writing projects, you begin to think about what has worked in the past, and you have a good idea of where to start.

Here is a typical situation: A technical communicator we'll call Heather is part of a technical communication group in a Midwestern manufacturing company. She is the safety regulations writer of materials that are produced for in-house use by employees who work in the shops and on the assembly lines. She also writes proposals for changing operating procedures within the company, reports to the state and federal agencies that require safety compliance and hazard information that is included in manuals used by workers in the plant where she works and in other sites operated by the company. In response to a change in Occupational Safety and Health Administration (OSHA) regulations, Heather has been given the task of updating the hazard and safety information in a training manual used by all employees at every site, so that the new edition of the manual will be in compliance with the new regulations.

Where does Heather start? She knows what kind of information appears in the old manual, but she needs to determine what needs to be changed. When she updated the previous version of this manual, she talked with a state employee who works with OSHA and knows all about regulations. She also worked with two supervisors on the assembly lines, who told her what she needed to know about current operating procedures and how the changes in regulations would affect the work flow on the lines. These supervisors also made arrangements for her to observe the ways employees did their jobs. Finally, Heather also made frequent visits to the various manufacturing areas within the company to see what kinds of safety information appeared in posters, as notices in lunchrooms, and on the equipment itself.

Heather knows the audience, the working environment, and the need for the information. She can begin to make preliminary notes, which can turn into new procedures. Her thinking might be directed as found in Figure 2.1, she might jot down or just keep in mind these topics as she plans her research strategy.

By comparing previous projects with the current one, Heather has a good idea of where to start. She knows she'll set up interviews with the state employee and the two supervisors, review the old manual, and observe employees and their work environment. Of course, that won't be the extent of Heather's research, but she has a starting point. As she has thought about her research objectives, she filled in her chart as shown in Figure 2.2.

You may want to create an objectives chart similar to Heather's, or you may just need to review mentally this kind of information. But always start with what you know. Then you can determine how much you still need to learn.

Review What You've Recently Learned

Technical communicators must keep up with changes in technology, science, and information-creation processes that affect the way they work,

What I Know	What I Need to Discover or Verify	Possible Sources of Information	What Each Source Will Provide

Figure 2.1 Heather's research objectives.

What I Know	What I Need to Discover or Verify	Possible Sources of Information	What Each Source Will Provide
My audience	New regulations	State employee	New regulations
Their environment Company specs	What needs to be changed in the old manual	Supervisors	Information about the effects of the new regulations
Company's need for changing procedures	How the information should be designed	Old manual	Information to be updated
What I've previously written about OSHA regulations	How the information will affect employees and the company	Other OSHA-related information	Information to be updated
What I've previously written in other manuals		Observations in the plant	Locations and types of OSHA-related information

Figure 2.2 Heather's research objectives.

even though those changes occur continuously. Fortunately, there are professional development activities for technical communicators, such as conferences, workshops, seminars, and professional meetings. In addition, most technical communicators subscribe to at least one, if not several, professional journals, both in technical communication and in their technical or scientific area of specialization. Because online information and computer networks are becoming accessible to more individuals, it's also easier to keep up with multimedia information available about different subjects and to "talk" with colleagues via e-mail.

When you are given a topic about which you'll write, you review what you've recently read, heard, discussed, and seen. You may not be sure of all the facts, but perhaps you can recall—even vaguely—a trend or a new approach to the subject. You may have learned the name of someone who could be an SME for your project. When you review what you've recently learned, you don't have to be an expert on the subject. You just need to keep up with changes in your profession. By reading, listening, looking, and discussing, you may not have all the answers, but you've gathered ideas that you can later tap into. This is another good place to start, with those memories of what you know exists somewhere or what has recently changed. If you can capture some ideas about what is new or innovative, you can follow up when you have to go to sources of information.

For example, Heather, our example technical communicator in the manufacturing company, is a member of the Society for Technical Communication (STC) and so receives the journal, *Technical Communication*. It recently had an article about training that she remembers may be useful as she writes new information for the manual. It also noted some changes in hazard notices and provided a description of ways to write them effectively. Although she doesn't remember in which recent issue this information appeared, and she certainly doesn't remember all the specifics of how to improve the information she'll write, she remembers the general content of the articles and knows she wants to consult them before she begins to revise the safety manual. Heather also is a member of the American Society for Training and Development (ASTD), and at a recent meeting, the speaker discussed the issue of liability in writing safety materials and provided some tips for helping ensure workers' safety. Heather has the speaker's business card and plans to talk with this person as a possible new SME for her project. Thus, Heather's research outline of objectives may be expanded to include this new information. (See Figure 2.3.)

Taking stock of what you've recently learned is a good step toward research when you have a subject and a limited amount of time in which to conduct the research needed for writing. This inventory of information is also a starting point for further idea generation, because once you recall—and preferably jot down—a list of possible sources of information, you're more likely to brainstorm additional sources of information or possibilities of contacts or secondary sources that may be useful as you plan your research and writing tasks.

Getting Ideas for Your Projects

Research for its own sake is a valuable enterprise. Of course, this research often leads to practical applications in science, technology, communication processes, and information designs, but its purpose is more altruistic. When you want to conduct research to increase the body of knowledge about a subject, you need ideas to plan a research design and methods to complete the research. This process may take a long time, and the idea-generation process must be natural and ongoing. It's difficult, at best, to come up with a great idea under pressure, so you're more likely to

What I Know	What I Need to Discover or Verify	Possible Sources of Information	What Each Source Will Provide
My audience	New regulations	State employee	New regulations
Their environment	What needs to be changed in the old manual	Supervisors	Information about the effects of the new regulations
Company specs			
Company's need for changing procedures	How the information should be designed	Old manual	Information to be updated
What I've previously written about OSHA regulations	How the information will affect employees and the company	Other OSHA-related information	Information to be updated
What I've previously written in other manuals		Observations in the plant	Locations and types of OSHA-related information
		Technical Communication article	Training information and hazard notices
		ASTD member	Safety information

Figure 2.3 Heather's expanded research objectives.

develop an idea for research if your idea-generation processes are a natural part of your daily research and writing activities. To get started in this type of research situation, you work with three basic processes:

1. Observe the world around you.
2. Review what other researchers have done.
3. Anticipate what needs to be done next.

Observe the World Around You

Good researchers are good observers. They constantly are aware of their surroundings and the interaction of people, machines, animals, plants, the elements—in short, the whole environment. They may be more focused on that part of the environment that is their area of expertise, but the focus doesn't prevent them seeing their part of the environment in a larger context. For example, researchers in psychology may be more aware of what people are doing and saying, whereas researchers interested in air quality may be more aware of the morning haze as they go to work or the black smoke coming from the tailpipe of the car in front of them. Even if you're not consciously paying attention to all the details of life going on around you, you're aware of some details more than others. These nagging little pieces of daily life can prod you to ask some questions or become concerned about a situation.

Observations can be more carefully planned, even long before you might want to design an experiment. For example, a researcher we'll call William is interested in the way people use computers in education, especially which kinds of information designs are more effective with adults. To get some ideas, he visits several computer labs on a university campus and observes adults working with the equipment. He also plans to observe people being trained in different occupations, such as bank tellers, data entry operators, and accountants, to see how people are learning to work with computers. Although he's not officially conducting research, because his methods are still rather haphazard and unplanned, he's getting ideas that may lead to formal research about one part of computer-based information design.

What you observe is stored in your mind, but you can help the idea-generation process by keeping a diary or a log of notes recording your thoughts and informal observations. Even when these notes don't lead you to an idea for further research, they help you focus your observations and become more aware of your surroundings. They also help you get in the habit of making informal observations, as well as making contacts with people who can help you when you're ready to conduct formal research. For example, William developed contacts with personnel in different computer labs as he informally observed what takes place in the labs. These contacts will be useful if he wants to set up formal observations using lab patrons or the computer lab as a test site.

Review What Other Researchers Have Done

As a technical communicator and a researcher, you may be familiar with several scholarly publications. You might subscribe to or regularly read in the library several research-oriented publications. Most of these publications are still in paper form, although some also may be available electronically. You probably are familiar with online bibliographic services and networks that provide you access to new information being distributed to professionals in your area of expertise. You may also read research newsletters and bulletins to keep up with what others believe needs to be researched and what is currently being done in your field.

None of us can or will keep up with all the publications in our areas of interest, but researchers need to know what other professionals are reading and what they may know from sources we don't normally peruse. Library research therefore becomes an important part of professionals' idea-generation strategies, because secondary sources are useful in these ways:

- They indicate how much has been published or distributed about a specific topic.
- They list the names of researchers who are working within a specific field or are conducting certain types of research.
- They indicate trends in research methods and subjects.
- They provide sources for bibliographies and literature reviews.
- They describe and explain new information that's an important part of the common body of information about a subject.

To learn what others are doing in your area of expertise, you may need to work with these secondary sources: newspapers, government periodicals, abstracts, newsletters, journals, and magazines. Then, if you're interested in becoming expert in a subject area or if you just want to learn more about a particular subject, you may need to refer to handbooks, encyclopedias, dictionaries, indices, and other reference books.

As you review what other researchers have done, you develop ideas for your research. For example, you may follow the publications of one or two prominent researchers and decide to replicate their work. Or you may read about several experiments along similar lines and develop your own experiment to test a hypothesis. Or you may wonder what would happen if you changed a research variable, for example, the number of participants involved in a project. Or you may discover what you believe is a flaw or a misconception in someone else's research design and think of a way to improve it. Reviewing others' work can provide you with specific ideas for your research.

Being aware of what others in your profession are doing also alerts you to trends and current research interests. This information may prompt your thinking about research projects along similar lines. It can also indicate research areas that are not getting the attention you think they need and may provide a fruitful area for your consideration.

Anticipate What Needs to Be Done Next

Through observation or reading, you may have discovered something that interests you but is not currently being researched. Or you may be very familiar with what has been researched recently and understand the next logical step for that research. In either case, you anticipate what needs to be done next.

Some technical communicators work very practically with equipment, such as robots or computers, and know the current state of the technology. They envision what they would like to see happen "someday" in terms of practical improvements in design and function. Although the technology may not currently be ready for that leap to "someday," these visionaries anticipate what needs to be done and start conducting research to move technology in that direction. Through a series of small projects, with the objective of practical improvements to equipment, these researchers help turn what should be into what is.

Other researchers prefer to consider theory, without needing to anticipate a direct application for their research. They have an idea of how something works, but they have been unable to prove it. They anticipate cause-and-effect relationships, for example, but lack direct evidence that these relationships do, in fact, exist. Theory and paradigm development are important areas of research, and your familiarity with a subject may start you thinking about what you—and other researchers—don't know but should investigate.

You should understand, however, that you can't always anticipate research. Sometimes serendipity plays a big part in what you learn or ultimately decide to research. Although the research methodology must be carefully planned, inspiration can't be produced on demand. Nevertheless, if you keep observing, reading, discussing, and thinking, you're more likely to become inspired.

Idea Generation with Computers

Computer technology is such an integral part of our research methods throughout the research process that we often take it for granted when we gather information, record results, analyze data, and share information. But computers can also help you generate ideas, from a very simple use of word processors, to database creation and research, to experimentation with computer technology. The widespread use of multimedia and

CD-ROM technologies, for example, has helped make computers even more effective in idea generation, as well as in other research-related tasks. For whatever type of research you're required to do or would like to do, computer technology can help you with every phase of that research.

Word Processors in Idea Generation

Using word processing to generate ideas is practical because it is so efficient. It allows you to concentrate on ideas, rather than on structuring the ideas, because files can be easily organized and reorganized.

Three strategies you may have learned in a composition class to help you overcome writer's block also work when you're generating ideas for research. *Brainstorming*, *outlining*, and *freewriting* are helpful when you need to generate a number of ideas and later evaluate them. Especially if you have a deadline for writing information, the idea-generation part of your writing process may need to be completed quickly. If you're thinking about a subject and trying to get a focus for research you might design, these idea-generation activities can also help you open your mind to new ideas and record a wide range of ideas that might prove useful later.

Brainstorming

Brainstorming is one of the most frequently used idea-generation activities. In a brainstorming session, you let your mind wander around the subject about which you want to generate ideas. Then, as quickly as possible, you record the words or phrases that come to mind. This type of free association can generate many ideas very quickly, and the brainstorming session usually lasts only a few minutes. The purpose is to generate as many ideas as possible within a short time. After the brainstorming session, you evaluate the ideas, deciding which have merit for further brainstorming and which are dead ends, at least for this project. You then reorganize and develop the idea(s) that seem(s) promising.

Brainstorming is particularly well suited to word processing. When you get an idea, want to record a question, or have a topic about which you need to brainstorm, you use the keyboard to create your information. You can very quickly key in words and phrases and, at the end of the brainstorming session, use cut and paste, find or replace, and sort features to help you organize the thoughts that seem worth pursuing. The results of your brainstorming become an idea file. An idea file, in which you store your brainstorming sessions, notes, questions, and descriptions of observations, is a handy tool that can be used either in word processed form or can be turned into database entries that can be sorted and cross-referenced.

In their simplest form, the screens of information in the file can be developed into outlines, or you can cut and paste the information from the file into a document as it is being developed. The file is stored for future reference, and you will probably add many other screens of information to this file as you develop ideas during other brainstorming sessions.

For example, a technical communicator named Raye needed to brainstorm the topic of virtual reality to get some ideas that may be further developed into a conference presentation. The information she developed with a word processor during a three-minute brainstorming session is shown in Figure 2.4.

misused term today
needs a better definition
almost like reality
holograms
Star Trek
games
computer pinball
dungeons and wizards
helmets
visors
gloves
3-D glasses
need for being in a net environment without goggles
great way to learn
medical uses
news magazine articles about medical students using VR to do operations on
  computer-generated patients—need to find that article or those articles
architectural uses
build a house mentally instead of "really"
expense?
home uses today or far into the future?
fun uses
will it mess up someone's understanding of a real reality?
can it be bad?

Figure 2.4 Notes from a word-processed brainstorming session on virtual reality.

This brainstorming session provided information about several different aspects of virtual reality that could be worked into a conference presentation. The information later was sorted electronically using search and sort functions to create the lists shown in Figure 2.5.

Virtual Reality

Basic Information about Virtual Reality
- misused term today
- needs a better definition
- almost like reality

Devices Needed for Creating Virtual Reality
- helmets
- visors
- gloves
- 3-D glasses
- need for being in a net environment without goggles

Uses for Virtual Reality
- holograms
- *Star Trek*
- games
- computer pinball
- dungeons and wizards
- great way to learn
- medical uses
- architectural uses
- build a house mentally instead of "really"

Examples of Virtual Reality Applications
- holograms
- *Star Trek*
- games
- computer pinball
- dungeons and wizards
- news magazine articles about medical students using VR to do operations on computer-generated patients—need to find that article or those articles
- fun uses

Practical Considerations for Virtual Reality's Uses
- expense?
- home uses today or far into the future?
- can it be bad?

(continued)

Figure 2.5 Word-processed lists created from the original brainstorming session.

Ethical Implications of Virtual Reality
will it mess up someone's understanding of a real reality?
can it be bad?

Areas for Further Research
news magazine articles about medical students using VR to do operations on computer-generated patients—need to find that article or those articles
expense?
home uses today or far into the future?
will it mess up someone's understanding of a real reality?
can it be bad?

Figure 2.5 Word-processed lists created from the original brainstorming session. (*Continued*)

Once you have this kind of list, you add notes or additional questions about the ideas you generated. You might decide to brainstorm on another word or one of the key areas that was developed from the original brainstorming session. When you save this idea file, you have a starting place for several different approaches to researching the topic *virtual reality*.

Outlining

Another important idea-generation activity is outlining. If you tend to group ideas or easily create lists, you most likely will find an outline useful as you think about a topic. Most word processing software comes with an outline feature. When you use this feature, you can jot down topics you have been instructed to research or that you would like to research. Then, as you come up with ideas, you can add points and subpoints to the outline. The software automatically renumbers the ideas and expands to fit your expanding idea. The information in the outline can be deleted, revised, or reorganized quickly and easily with the outline feature, and you can immediately get a sense of the parameters of your project or the research you'd like to design. This outline can later be revised as a table of contents or an approach to conducting research.

A sample outline of ideas generated about virtual reality was developed using a word processing package's outline feature. The final outline, after the ideas were reorganized, further developed, and then automatically reformatted and renumbered, looked like Figure 2.6.

An outline feature usually offers several formats for the outline, from the traditional Roman numeral style to legal numbering systems (e.g., 1.0,

I. Current definitions of virtual reality (VR)
  A. Do not provide an accurate definition
  B. Often reflect applications, not the entire scope of VR
  C. Need to be distinguished from simulation, for example
II. Popular applications currently available
  A. Computer games
  B. Educational software
  C. Training simulations
III. Positive implications of VR
  A. Better educational and training methods
  B. Cost-effective modeling and planning
  C. Experiential learning and development
IV. Negative implications of VR
  A. Ethical considerations of tampering with "reality"
  B. Cost considerations for making VR available to everyone

Figure 2.6 A word-processed outlining session on virtual reality.

1.1, 1.1.1, 1.1.1.1) to bulleted styles (e.g., •, ○, *) to styles you design. The information created in the outline is turned into regular word-processed text once the outline feature has been turned off.

Freewriting

A third way of generating ideas using a word processor is to freewrite. Freewriting is a process of continuous writing for a limited time. You may want to set a timer or an alarm for three to five minutes, so that you limit the time you're writing. Once you begin freewriting, don't let your fingers leave the keyboard for the allotted time. Even if you key in the same word or phrase over and over, or if you only key in "I don't have any ideas right now," keep writing. The purpose of this activity is to get some information on the screen, even if only one or two words end up being important. Chances are, as you settle into writing, your thoughts turn to the subject of research, and you capture at least one or two ideas that are worth following up. Again, the highlight, delete, move, and copy functions will be most useful as you evaluate what you wrote during the freewriting session. The pared information can be saved as part of an idea file or developed into an outline.

The freewriting sample found in Figure 2.7 has not been revised or edited, although the technical communicator who created it can later return to the file to evaluate what was created and to highlight important

What does virtual reality really mean? How can something real be "virtual"? I don't know much about this subject. It seems overwhelming right now—but I guess that's why I need to clear my mind and figure out what I want to do about it or what I need to learn about it. OK—where do I start? Science fiction—thats a good place for virtual reality. the characters on TV use VR when they want to talk to the computer and create a playroom—that's not what they call it. How about when we play a game and get really involved with it? Is that VR when we suspend belief or is it just a good game? All games aren't VR, are they? Only computer games? What about CD-ROM? Or how about using a visor to see things that aren't there and to react to them? If we're liited by space and time, how can it be a new reality? What if I believe there's a cabinet in front of me because I'm wearing a visor, and the visor makes me see the cabinet, but it's really not there and I try to walk through it? That probably would make me believe in the real reality, not what I was seeing through the visor. Is that what makes it virtual? I don't have very many answers, do I? But I sure have a lot of questions. But maybe there are no answers now. Perhaps we should see what develops over the next few years, or is developed by scientists and otehrs, and in the interim hope big brother isn't watching too closely. Vr is good subject for movies, but if that's true then it should be good vehicle for training uses. Why nothave a mechanic in a shop watch someone repairing s motor on vr while she doesthe same work on th car. And now my time's up, so I guess I'll have to do this again later.

Figure 2.7 A word-processed freewriting session.

information or delete unnecessary information. Few ideas generated during the freewriting session may be profound, but at least one or two good ideas usually result from a short freewriting session.

The technical communicator who created this information during a three-minute freewriting session was a good writer who was very familiar with the word processor. Most freewriting sessions result in text with considerably more typographical errors and fragmented ideas than this text contains. As you freewrite online, you concentrate on letting your mind and fingers work together to key in ideas on the screen. In fact, you may want to look away from the screen, close your eyes, or turn down the brightness knob so the screen is dark. That way you eliminate the distractions that can break your concentration on the topic.

After you've taken a break from freewriting, you eventually evaluate the information you created. Figure 2.7 was revised later to create a word-processed idea file consisting of questions that the researcher later reviewed and used as a starting point to learn more about virtual reality. (See Figure 2.8.)

What does virtual reality mean? How can something real be "virtual"? Virtual reality might come from science fiction and the VR applications shown on popular television programs. Note: Find out more about current television programs and movies using VR. Are all computer games examples of VR? What about using a visor to see things that aren't there and to react to them? If we're limited by space and time in any environment, how can a visor create a new reality? For example, I believe there's a cabinet in front of me because I'm wearing a visor, and the visor makes me see the cabinet. But it's really not there and I try to walk through it. That probably would make me believe in the real reality of walking through a space without the cabinet, which is not what I was seeing through the visor. Is that technology what makes reality virtual?

Figure 2.8 A revised word-processed freewriting sample.

One idea in particular, the cabinet example, seems to be more promising and might be one that should be explored, either through another freewriting session or with another technique for generating ideas.

Graphics Software in Idea Generation

Some technical writers are really more visually oriented and need to see their ideas in graphic form rather than in text. Graphics software can provide an easy way to draw instead of key in ideas.

If you're working with boilerplated graphics like blueprints, and brainstorming new designs or ideas that may eventually work into a better design, you can manipulate the boilerplate with keys, a mouse, or a light pen, for example. You can zoom in and modify only one part of the graphic or zoom out for the big picture.

If you're using a CAD model, you can rotate, lengthen, shorten, explode, compress, and probably select colors for your design. At the press of a key or two, you can change your selection to view the information in another way.

If you like the idea of brainstorming with graphics, you may use a drawing software to create a flowchart or diagram of your idea, or you may draw a design for a product you envision.

Graphics software let you work with existing graphics or graphics you create. As in using other idea-generation activities, you want to develop as many visual ideas as possible within a short time. Later, you modify your design, delete or insert segments, and enhance the basic concept.

Bibliographic Software in Idea Generation

When you work with word processing, database, or bibliographic software, you create a working bibliography of sources to which you may turn for more information or for inspiration. As you add to an existing bibliography, you generate ideas by perusing the entries or completing keyword sorts and searches to locate sources about similar topics or written by the same researcher.

Word processors provide a simple way to generate ideas through bibliographies. As you gather information, either by keying in bibliographic citations or downloading them from other bibliographies, you create your own bibliographic file. When you need to get some ideas, you use the word processor's find or sort commands to locate individual sources or group them in new ways that can stimulate new ideas.

Databases provide a better means of recording bibliographic information. As you develop your database, you arrange the fields and records in ways that will make sorting the records and searching for similar fields easier and better suited to your research needs. Database software can help you group bibliographic information in different ways, which can lead you to a new approach to your research or to an awareness of the amount and kinds of information available about the subject you'd like to research.

Bibliographic software provides you with a consistent method of formatting and organizing information according to standardized citation formats. Some software lets you format bibliographic information according to American Psychological Association (APA) and Modern Language Association (MLA) specifications. If you need to reformat the basic information contained in the entries, you simply select another citation style, and the entries are automatically reformatted to meet the new specifications. Bibliographic software makes it easy for you to find information created by prominent researchers or organized by publication date, which can help you determine which sources you should locate or what needs to be researched.

Other chapters in this book provide you with more information about online bibliographies. You may want to look at Chapter 5, Computerized Information Retrieval from Libraries, Chapter 6, Computerized Secondary Sources of Information, and Chapter 16, Bibliographies and the Documentation of Sources for additional information about bibliographic citations, interactive software, and networked databases.

Interactive Software in Idea Generation

Several types of interactive programs help users record and track their ideas as they're developed or allow them to consider different associations among ideas. Interactive computer information is common in a variety of settings. It is useful in tutorials, in which users respond to prompts and get feedback about their responses. It's also used in educational settings to help students develop their ideas through a series of computer-generated questions and user-written responses. Through hypermedia or hypertext software, users have been prompted to make associations among topics, which they might not make without computer prompting.

Interactive Software, Multimedia, Hypermedia, CD-ROM, and Virtual Reality in Idea Generation

Interactive software may be programmed through authoring languages, for example, to create a very specific program and involve limited interaction through a keyboard or a mouse. Students or researchers might answer questions and develop associations among concepts in response to computer prompts. Teachers or trainers often customize the software to meet their classes' needs. This type of program is quite effective, and it is relatively simple to work with.

As computer technology has become more sophisticated, so have the interactive capabilities of many programs. *Multimedia* provides users with several forms of information, including written text, sound cues, music, videos, and animation. Multimedia gives users the option of seeing only one or several formats of information alone or in combination, but the information is often arranged linearly, or in a predetermined sequence.

Hypermedia adds another layer of interaction by allowing users more choice in selecting the amount and type of information they want and the sequence in which they want to access it. Although purely random organization of information is not currently possible, hypermedia provides users with the ability to select information according to the associations they make among topics. For example, one researcher may want to learn more about toxic wastestreams. He or she may choose to search for information about hazardous materials, then remediation techniques, then the equipment needed for such treatment. Another researcher, also interested in toxic wastestreams, only wants to know what kinds of toxic wastestreams have been found near his or her hometown in the past six months. The first researcher may prefer to read the information, which may include government reports and encyclopedic descriptions of remediation processes, whereas the second researcher wants photographs and maps of the waste sites. These types of information can be easily accessed through hypermedia.

CD-ROM provides this multimedia and hypermedia information on disk. Users can access the information, interact with it, stop it, download it, and perform other tasks as necessary to meet their research needs. Although the information on CD-ROM may become outdated, and the cost of buying new disks periodically can increase research costs, CD-ROM is a valuable research tool that technical communicators cannot afford to ignore.

Virtual reality may be computer generated or created through other devices, such as helmets, hoods, or visors. Virtual reality simulations can provide you with experiences based on the way you interact with and react to the environment that you have helped create. Especially with virtual reality, which in its sophisticated forms "tricks" the mind into believing you are in a different environment, you test your ideas or experiment with different activities and concepts. By working in a virtual environment, you safely determine what you can or cannot do, what should or should not occur, and what results when you follow a certain sequence of activities. In addition to being a good learning or training tool, simulations can give you ideas about what can be tried next and provide a safe way for you to test your ideas.

Although multimedia, hypermedia, and CD-ROM technologies are not available to all technical communicators, they are becoming more commonplace in research and have found their way into popular entertainment and educational venues. If you can develop databases, especially in hypermedia, you can generate ideas in a number of ways. By searching for keywords as they come to mind—similar to brainstorming—and selecting different ways to use the information (through sight, sound, or touch, for example), you can generate other ideas about your research topic. Virtual reality simulations may provide you a computerless way to generate ideas or may help you, through computer-based simulations, develop or test your research ideas. The key to using these tools in idea generation is to make the technology useful to you and the way(s) that best help you to get as many ideas as possible and record those ideas for current or future use. Throughout this book, you'll find examples of some of these technologies and read about additional ways that computer-based technologies can be used throughout the research process.

Other Computerized Sources of Information

In this section, you've read about several ways you can use the computer to generate ideas, which in turn become computerized sources of information when you next need to begin planning research. These personalized sources of computer-based information become valuable to you as you continue your work as a researcher, because in effect you're developing your own networks and databases for future use.

In addition to computer use in idea generation, you also should be familiar with online sources of information and ways to gather the information that has already been compiled by other researchers. In Chapters 5 and 6, you'll read how to use the computer to locate secondary sources of information, as well as to create primary sources of information.

Summary

Whether your need for research ideas is practical and designed to help you complete an assigned project or is ongoing and tuned to research you'll design later, you need several strategies for generating ideas. You may find that reading, observing, and doing other tasks in the "real world" help you generate ideas and make you aware of what should be researched and how you might approach gathering information. Or you may prefer to work with computers exclusively, not only to help you generate ideas but to record those ideas you get. However you generate ideas, as a technical communicator you need to be creative, innovative, and yet practical as you develop ideas for your research. The key to good research is not only getting ideas, but also finding ways to record them during periods of inspiration, and then developing them into workable forms for more research or writing.

Research Activities

1. Look through recent issues of *Technical Communication*, *Journal of Business and Technical Communication*, *Technical Writing Quarterly*, and other technical communication journals. Then answer the following questions:

- What are current trends in research in technical communication?
- Where are bibliographies that can provide you with sources for your own research?
- Who are leading researchers in technical communication?

The answers to these questions change as research is advanced, so you may want to conduct this exercise monthly or at least quarterly to keep up with new trends and topics in technical communication, as well as within your area of specialization.

2. Using a word processor, list your research interests in separate files. Select one subject that especially interests you now. Then brainstorm for five minutes about this topic, just to see what ideas you have and where these ideas might lead you. Save your brainstorming session under a separate filename so you can return to it, as well as your files of research interests.
3. Check with faculty, staff, or personnel in computer labs, libraries, computer services, and computer sciences departments in universities, communities, and companies in your area to learn how to access interactive computer programs and idea-generation and development software. Use at least one program to help you generate ideas for a current research project.
4. Begin developing an online bibliography of sources you find about your research interests. Work with a database, word processing, or bibliographic software to create an ongoing bibliography for each area of research that interests you.

For Further Reading

Barker, Thomas T. 1989. "Word Processors and Invention in Technical Writing." *The Technical Writing Teacher* (Spring): 126-35.

Desai, Pranav Kiritjumar. 1994. *Brainstorming as a Concept Generation Methodology in Mechanical Design.* Master's thesis. Case Western Reserve University.

Elbow, Peter. 1992. "Freewriting and the Problem of Wheat and Tares." *Writing and Publishing for Academic Authors,* pp. 33-47. Joseph M. Moxley, ed. Lanham, MD: Univ. Press of America.

Goss, Larry D. 1987. "Techniques for Generating Objects in a Three-Dimensional CAD System." *Engineering Design Graphics Journal* (Fall): 29-35.

Green, W. T., L. V. Sadler, and E. W. Sadler. 1985. "Diagrammatic Writing Using Word Processing: Computer-Assisted Composition for the Development of Writing Skills." *Computing Teacher* (April): 62-4.

Jonassen, David H., and Heinz Mandl, Eds. 1990. *Designing Hypermedia for Learning.* Series F: Computer and Systems Sciences, Vol. 67. Berlin: Springer-Verlag.

Mauer, Mary E. 1994. "Brainstorming." *Writer* (December): 22-3.

McKnight, Cliff, Andrew Dillon, and John Richardson. 1991. *Hypertext in Context.* Cambridge: Cambridge UP.

Nadis, Steve. 1994. "Brainstorming Software: Can Your Computer Be Equipped with Artificial Creativity?" *Omni* (December): 28.

Richards, Thomas C., and Jeannette Fukuzawa. 1989. "A Checklist for Evaluation of Courseware Authoring Systems." *Educational Technology* (October): 24-9.

Robbins, R. 1987. "Helping to Make Reports Real: A Brainstorming Aid for Assignments in Technical Communication." *The Technical Writing Teacher* (Spring): 99-102.

Sugg, David. 1993. "Putting Meat on the Bones—Cause-and-Effect Diagrams." *Plating and Surface Finishing* (November): 54-5.

CHAPTER THREE

Planning Research: Documentation Plans

The complete process of taking an idea through multiple stages, from concept to presentation, is characterized by two very general types of tasks: information-management tasks and resource-management tasks. In the category of information mangement, we might include, for example, research, writing, editing, reviewing, testing, developing graphics, and producing the final document. These kinds of tasks concern developing the content, design, and format of the information.

Every task involved in information development, particularly in a business environment, is affected by the resources available to complete the tasks. For example, projects must be completed within specific time lines, and costs must stay within a budget. Furthermore, all tasks may be the responsibility of one person, or they may be the responsibility of several people, each of whom may be working on more than one project simultaneously. Various tasks may require specific equipment or facilities. Every task will certainly require time and money, and for most, if not all tasks, the process and the results of completing the process will be governed by standards, specifications, and policies. Although it may be somewhat difficult to call budgets or specifications resources, they do, in fact, provide important guidelines for developing documents.

The key to successfully completing any project, then, is to recognize all of a project's requirements, identify the tasks necessary to complete those projects, and plan them. Developing informational products is, therefore, a management process. In this chapter, we discuss strategies for focusing your ideas to help define your research objectives through planning document(s) and identifying and planning your use of resources.

Doc Plans

If you work in an industry, when you are given an assignment, you might get a definition of the type of document(s) you'll create; or, because of your experience with similar assignments, you may know the kind of document you need or want to create. At other times, however, you may have an idea you want to research, but you have no clear vision of the final products that may result from your research. In either case, one principle focuses all work: people want or need the information.

All documents are designed to provide audiences with information that they want or need to know or use. When they use it, they use it with their own objectives in mind and in a particular environment. A *doc plan* helps provide you with a blueprint for your tasks, from the initial research through the final distribution of the product, to ensure that your document meets its objectives, and therefore will help your audience meet their objectives.

The activities that are described in a doc plan help you define the document and the tasks that you (and probably others) will have to do as the document is prepared. Those tasks include first focusing on the desired result. To develop a doc plan, you'll need to do the following:

- Define your objectives.
- Define the document.
- Outline the content.
- Define the research.
- Determine the project's limitations or parameters.
- Plan the management responsibilities.
- Design the information.

- Plan the visuals or special effects.
- Produce the document.

Each of these tasks can be further divided into subtasks and assigned to a person or a group. The outline of who is responsible for what within a time frame will become the formal doc plan.

Define Your Objectives

Before you can know what to research, much less to write, you need to define your objectives for this project. To do that, you ask the following questions:

- Who is the audience?
- What information do they need?
- What will they be doing with the information?
- In what kind of environment will they be working?
- What product, service, or process are you documenting?

Define the Document

After you've developed objectives for the information, you begin to define the "document," or whatever form your information will take. (*Document* refers to any written or designed technical or scientific information in any medium. A document as we're using the term does not have to be on paper.) Documents are often named by type, according to the objectives they're designed to meet, the information they will provide the audience, and the audience's use of that information. The type also defines a format and a structure, as well as a plan of development. For example, if you're asked to characterize an instructional manual, you might describe it as a set of instructions designed to tell an audience how to do a task. The manual begins with the first information the audience will need and provides the information step by step to the last information they need to complete the task. As a writer, therefore, who is assigned to write instructions for completing a task, you apply your understanding of the characteristics of instructions to help you determine some parts of the structure and content of your document.

In a working environment, you'll probably be told the kind of document you need to produce as part of your work assignments. But that is not true of all the documents you might produce. For example, if you write an article for a journal, you might have to deduce from reading other kinds of articles in the journal what the format, structure, and general style should be. And you compare those elements with the information provided in Information for Contributors or other similar guidelines.

You also define the document through your understanding of common structures for hardcopy or softcopy information. This understanding may cause you to redefine your original concept of the document. Paper, or hardcopy, documents are by necessity and tradition linear structures of information that have readily definable parts—beginning, middle, end. In addition to these definable parts, there might be prefatory material or appended material, but even the words *prefatory* and *appended* support the idea that information is presented linearly—something must come before something, and something must come after something. The structure you plan for a paper document adheres to the conventions of linear documents. You plan a table of contents as prefatory matter to let readers know where information is located and what kinds of information are in the document. You also plan the order of major sections, such as chapters, in a linear order from general to specific information, easiest to most difficult, first in a sequence to last in a sequence of activities or events, or another hierarchy of information. You assume that readers will move from beginning to middle to end in the document. Your plan for the document's structure, format, and style reflects that assumption even though you add tables of contents and indexes, for example, should the audience desire to access the document in a nonlinear fashion.

Technical and scientific information is no longer tied to those kinds of artificial boundaries, because you're not limited to writing hardcopy documents. You may decide that the best format for your work is not a document at all, but an interactive disk. The order in which information will be presented may then consist of "chunks" of information that readers/users can work with in any order they choose. The definition of your document will be very different in this situation.

To define the document, ask these questions:

- Has the document type been clearly defined?
- Does the subject lend itself to presentation in one medium, format, or structure more than in another?
- What kinds of formats, media, and structures of information are best for your audience?

Outline the Content

Outlining the content of your document is really the beginning of a written doc plan. As your idea of the document takes shape, you can ask these questions:

- What is an outline of other similar documents, if a document type is assigned?
- If a document type was not assigned, are other similar structures available that you might consider?
- What are major categories of information you will need to provide in the document?
- Do these categories have a relationship to each other, which will define the order in which they're presented?
- How can you outline the information?
- Should the information be mapped?

As you outline or structure the information, you put in both information you know and information you think you'll need to know. The structuring process may create as many questions as answers, but throughout the process, you add those questions. Your design should describe how you envision the product, but both your vision and the design should be flexible. Documents generally turn out to be better products when the design accommodates the information and the audience's need for information, rather than the information accommodating the design.

Figure 3.1, for example, is an outline of content written by a person assigned to create an article introducing the ISO 9000 standard to a novice audience. The outline was intended to focus research, not to be necessarily the final outline for the paper.

Determine Research

Throughout any planning process, you should continually consider both the kinds of information you'll need and possible resources. To help you determine how much and what kind of background research you need to complete for this project, ask the following questions:

- What literature is available on the subject?
- Where can you find the literature? in databases? articles? boilerplate?

Assignment outline:

Introduction: Identify audience, subject, purpose, and scope of document

Audience: Novice to ISO 9000, but mostly professional technical writers. Educated or experienced in technical writing

Subject: Introduce ISO 9000

Purpose: Provide definition and show how standard affects the technical communication profession

Body

Definition of IS0 9000

Give definition of the standard

History, purpose

Explain what areas of industry ISO 9000 affects

How will affect technical communication

Give examples of possibe effects in a couple of situations

Explain how standard is implemented

Note possible areas of further research into the standard

Provide a reading list, list of organizations that provide ongoing information about the standard

Figure 3.1 An outline of an assignment.

- What is your experience/knowledge about the subject?
- Are there other people with whom you should talk?
- What kinds of information can they provide you?
- How should you make contact with other people?
- Can mailing lists or newsgroups provide information?

When you've determined possible sources of information, review your outline and determine which sources provide the best possible information for each section of the document. For example, for history and background, you might determine that boilerplate is the best source for long documents, like manuals. For other documents, you might decide that journal articles provide the best source of historical information.

For the person writing the article on ISO 9000, the best source of information on the history would probably be the standard itself.

To develop the document's content, you choose from a variety of sources. You may, for example, determine that the best information comes directly from developers or other subject matter experts (SMEs) for one subject matter, whereas for another you want to conduct some personal observations or experimentation. Your experience may provide some personalized content to which the audience can relate. However, personal experiences are most effective when they are generalized; thus, you may need to complete some research to discover others who have similar findings.

Define Resource-management Responsibilities

The document design you think would be best and the research you believe would be most helpful are ideals in your planning. However, technical communicators work in a real world, a world in which there are limitations. Some may be negative limitations, for example, time, but others can be positive limitations, for example, standards and policies that reflect uniform codes of conduct. As part of any planning situation, you identify those limitations or guidelines that provide parameters for your project. Although at this point you're really concerned about the research you'll need to do, before you write you're thinking about that research in terms of a final document. Thus, you should identify all limitations and parameters that affect all parts of the document-creation process. Ask these types of questions:

- What is the time frame for this project?
- Within the time frame, how much time can you devote to the project?
- Will other personnel be required?
- How much time will they need to devote?
- What is the project's budget?
- What facilities/equipment will be needed?
- Are there any lab or other space requirements?
- Will meetings with other people be necessary?
- When should those be scheduled?
- Do you need to schedule reviews? What kinds? By whom? How many?

- Do you need to schedule facilities or equipment use?
- Will you need to develop or schedule usability tests?
- How will you make budget decisions?

Creating a document, whether you are writing from concept or assignment, is not just an exercise in communication. It is an exercise in communication management. For shorter documents, you may need to do no more than manage your time for research and writing, working within an editor's publication guidelines. For other documents, particularly longer documents that you create on the job, the management tasks might be much more complex. They may include planning and scheduling facilities and equipment, reviews, and other personnel, for example. To ensure that you know what your responsibilities are to a project, you should have a check sheet of all possible requirements and responsibilities.

When you have made the decisions included in the preparations you've made thus far in planning your document, you have completed the following tasks:

- Determined a document design to meet your and your audience's objectives.
- Planned general research requirements and strategies.
- Considered management decisions.
- Considered limitations and parameters that will affect both the final document and your process in completing the required work on the document.

As you consider each decision, you've begun to limit your document and your processes in a number of general ways. At this point, you can begin to think in specific terms about the structure, style, and organization of the content of the document. You might think about this in terms of writing the information, but as you develop content, you'll discover that putting the document together is not just a matter of writing. It is also a matter of integration of graphics and perhaps other media to present the information.

Design the Information

Some questions to ask as you design the document are these:

- Which format will be most effective considering the audience's needs?
- Does the audience have language or stylistic preferences?

- Are there specifications or standards that affect the language?
- Will there be multiple drafts of the document?
- What are the due dates for the drafts?
- Who'll review the document's language for style?
- Who'll review the document's content for accuracy?
- Does the content lend itself to more visual than prose representation?
- Do markup languages' standards apply to the document?
- Who is responsible for determining standards?
- How will the standards affect document design?

As you consider the various parts of the document's design, one consideration is the relationships among your audience, the material, and the document structure, and how those relationships are best suited either to prose or graphics. Those decisions include not only the kinds of graphics to include, but also who'll create them, how they'll be created, and the kinds of limitations you have on creating those graphics.

Plan the Visuals and Other Special Effects

Graphics or other special visual effects require special planning. The following questions can help you with this planning:

- What types of visuals are most effective for this content?
- What kinds of visuals are most effective for the audience?
- Can you use existing graphics?
- What new graphics need to be created?
- Who will design the new graphics?
- How long will it take to prepare the graphics?
- What standards or specifications affect the graphics?
- Who'll determine required captions, labels, and documentation for the graphics?
- Who'll create the required captions, labels, and documentation for the graphics?

Produce the Document

All your decisions throughout these planning stages have concerned the development of the document. When that document is ready to go to final form, it will be produced in some manner. What will be involved in that production depends on the kind of document, the practices and procedures of those who produce the document, and the requirements for producing the document so that it can be disseminated in its most usable form for ready access by the audience. Finally, you're ready to produce, or to oversee the production, of the document and can consider how the production will occur. Questions at this stage may include the following:

- Who will produce the document?
- Who will coordinate the production process?
- If this is a hardcopy document, what are the print specs?
- What typographical decisions need to be made to make the font, typeface, highlighting, color, and so on, appropriate and usable?
- When will copy be prepared for proofreading?
- Who will be in charge of proofreading?
- When will final changes be made?
- Who will make final changes?
- How many copies will be produced?
- How will the information reach the audience?
- Who will disseminate the document?
- If the document is produced in soft copy, what are the screen specs?
- Will the document need to be tagged for electronic use? If so, who will tag the document?

As the person who is responsible for creating and managing the document, you must know early in the document-creation process what kinds of production decisions need to be made so that you can plan for those both in designing the document and scheduling the work so that the document is ready for production.

Creating a Formal Doc Plan

Good writing is a matter of good planning. All of your planning so far has led you to the point where you can now write a formal doc plan, something that will help you keep track of both information- and resource-management requirements. The doc plan helps keep you on track as a technical writer and researcher, and it describes each part of the project and indicates how long each phase takes. It helps ensure that nothing is overlooked and that each person participates actively in the project.

Using the Sample Documentation Plan

If you use the blank format provided in Figure 3.2, you will account for all the areas listed previously; plus, you won't have to create your own format. You may want to develop a variation of this format online as a database, with each item as a field name. Or you may want to create a hardcopy form that can be used for multiple projects.

Documentation Plan

Date Assigned:

Author(s):

Subject:

Tentative Title:

Purpose:

Scope:

Assumed Primary and Secondary Audiences:

- Lay audience/Novice users
- Executive/Managerial audience
- Operators
- Technicians
- Experts

Audience's Use of Information: (e.g., perform a task, make a decision, replicate an experiment)

- Personal Use
- Job-related Use
- Interest in Subject

Figure 3.2 A documentation plan outline.

Audience's Environment when Using the Information:

Audience's Pressures:

Audience's Special Needs:

Audience's Preferences for Style, Format, or Design:

Restrictions/Allowances Possible Based on Audience Analysis:

Tentative Format: (e.g., manual, handbook, procedure, brochure, pamphlet, report, article, script)

Mechanical Elements Needed for a Hardcopy Document:

- Title page
- Copyright
- Table of Contents
- List of Illustrations
- List of Symbols
- Glossary
- Documentation
- Appendices
- Indices
- Numbering systems
- Levels of headings
- Binding

Prose Elements Needed in a Hardcopy Document:

- Letter of Transmittal
- Abstracts (informative, descriptive)
- Preface

Figure 3.2 A documentation plan outline. (*Continued*)

Theoretical or historical background

Literature Review

Discussion

Summary

Conclusions

Recommendations

Cautions

Warnings

Hazards

Notes

Prose Elements Needed in Softcopy Information:

Help documentation

Other documentation

Prompts

Menus

Index

Graphical Elements:

Complex or Simple

New or Boilerplate

Computer-generated or manually created

Interactive or static

Tables (informal or formal)

Figures

Captions

Documentation

Figure 3.2 **(*Continued*)**

Location and purpose of graphics:

Style:

- Organizational pattern
- Paragraph lengths
- Sentence lengths and structure
- Vocabulary considerations

Style Manual:

Specialized Dictionaries:

Specialized Encyclopedias:

Reviews Needed:

- Legal
- Marketing
- Engineering
- Research and Development
- Training
- Editing
- Consumer/Audience
- Validation
- Usability testing

Production:

- Type of production (e.g., saddle stitched binding, stapled pages, videotape, CD-ROM)
- Time required
- Cost (total and per unit)

Sources of Information: (Mark Primary and Secondary sources of information.)

- Personal observation

Figure 3.2 A documentation plan outline. (*Continued*)

Personal experimentation

Interviews (and with whom)

Questionnaires (respondents and sampling method)

Reference works

Periodicals

Software

Models

In-house publications

Other:

Documentation of These Sources:

Timetable for Project:

Information gathering

Organizing

Writing a first draft

Getting reviews

Writing a second draft

Getting reviews

Editing the draft

Validating the document

Usability testing the document

Getting post-validation reviews

Incorporating changes

Producing the document

Outline of the Project (tentative):

Project Distribution Date:

Figure 3.2 **(*Continued*)**

When you're creating the doc plan, you have to see how all the individual tasks—for you and everyone else—fit into one plan with one final deadline. The due dates and personal timelines for each part of the task need to be determined after the master doc plan has been finished, and you may need to adjust your "ideal" time frame for each task to fit a realistic, and usually demanding, schedule.

A good way to plan a project is to "back out" the schedule from the ending date. When you start planning deadlines leading to the final deadline—the distribution date—you list that distribution date first. Then you look at the task immediately preceding—production. You determine how long it will take for production; then assign a tentative deadline for the tasks immediately before production, one that will give enough time to complete all the tasks. You move backward through your entire list of tasks in this way until you arrive at the beginning of your doc plan—the starting date. If your backed-out schedule indicates that you need more time than is allotted for the project, you know you have some serious negotiating to do. Usually you have to compromise among everyone else's schedule—including all the other projects you and they have to do, the have-to-have-it-done date for the current project, and a reasonable amount of time to complete each task within the current project. You may not have all the time you'd ideally like to complete each task, but by the time you've set the schedule in the doc plan, you'll have a list of due dates for individual tasks within the project and an idea of how long the project will take. You and everyone else involved in the project can also begin to plan how it will fit in with the other projects that you're expected to complete on time.

The doc plan in Figure 3.3 is the result of numerous negotiations about schedule and budget, but it ultimately details the final plan to which everyone must refer.

For this project, the information-gathering phase, an important part of the research, took only a short time—one week. The SMEs (S. Jackson and D. Davis, primarily) provided written summaries for the writers, who then compiled and revised them for the final report. However, the research phase of this project was ongoing, as questions about the summaries needed to be answered, new graphics were produced and checked, and City representatives viewed drafts and added new information. Although the formal information-gathering period was fairly short and much of the necessary information was provided, the writers still needed to ask questions of their SMEs throughout the project.

Documentation Plan

Date Assigned:	April 20, 1994
Author(s):	Steve Owens—lead technical writer Lynda Mason—technical writer, layout specialist
Subject:	City Report for 1993
Tentative Title:	The City's Annual Report, January 1-December 31, 1993
Purpose:	To provide a record of the City's departments' accounts and activities for the Municipal Board and the public
Scope:	An 8-1/2" x 11" bound report with a 1- to 3-page report from each department

Assumed Primary and Secondary Audiences:

Lay audience/Novice users	Possibly the City's citizens
Executive/Managerial audience	Municipal Board members
Operators	
Technicians	
Experts	

Audience's Use of Information: (e.g., perform a task, make a decision, replicate an experiment)

Personal Use	Citizens may want to find out some information about the City or a particular department.
Job-related Use	Required reading for Municipal Board members
Interest in Subject	Municipal Board members are elected to their positions and need to be familiar with the City's fiscal and personnel policies, activities, and problems. Only very interested citizens with a vested personal interest in some aspect of the City's operation will read the report.

Audience's Environment when Using the Information:

Municipal Board members will have a copy sent to their homes, where they can

Figure 3.3 A sample documentation plan.

read the report at their leisure. The information may be discussed or referred to in future Board meetings.

Audience's Pressures:

Board members need to read and understand the report, but they are under no time or job pressures to read it immediately. If they are going to know what's going on in the City, they should be familiar with the report, but it's only one of several documents the Board uses throughout the year.

Audience's Special Needs:

Readers prefer a large, spiral bound report with a serif font large enough for easy reading.

Audience's Preferences for Style, Format, or Design:

A document that can stay open on a flat table for reference during meetings is preferred. As noted previously, the font should be serif and large enough to be easily read.

Restrictions/Allowances Possible Based on Audience Analysis:

Tentative Format: (e.g., manual, handbook, procedure, brochure, pamphlet, report, article, script)

Spiral bound report with 8-1/2" x 11" pages. Glare-proof white paper. Black print using 12-point Times. Lots of white space. Two levels of headings:

Level 1 Heading

Level 2 Heading

Clear reproductions of graphics provided by individual departments.

Mechanical Elements Needed for a Hardcopy Document:

Title page	Yes, with State seal and City motto boilerplated from last year's report
Copyright	Yes. See Dana Davis for forms.
Table of Contents	Yes. Both levels of headings
List of Illustrations	Yes
List of Symbols	No
Glossary	No

Figure 3.3 A sample documentation plan. (*Continued*)

Documentation	Sources cited by departments are included.
Appendices	No
Indices	No
Numbering systems	Legal system for graphics within sections, Roman numerals for section numbers
Levels of headings	Two, as noted previously
Binding	Spiral binding

Prose Elements Needed in a Hardcopy Document:

Letter of Transmittal	No
Abstracts (informative, descriptive)	No
Preface	Yes. See Randi Carson, who will write it.
Theoretical or historical background	No
Literature Review	No
Discussion	No
Summary	Yes, provided by each department
Conclusions	No
Recommendations	No
Cautions	No
Warnings	No
Hazards	No
Notes	No

Prose Elements Needed in Softcopy Information: N/A—hardcopy document only

Help documentation

Other documentation

Prompts

Menus

Figure 3.3 ***(Continued)***

Index	
Graphical Elements:	
Complex or Simple	Very simple graphs and tables
New or Boilerplate	All new, provided by departments
Computer-generated or manually created	Both, depending upon the department
Interactive or static	Static
Tables (informal or formal)	Approximately 35, all formal
Figures	Approximately 40
Captions	To be created by S. Owens and L. Mason
Documentation	Provided by departments
Location and purpose of graphics:	At the end of each summary section representing a department
Style:	
Organizational pattern	Alphabetical order of sections by department name
Paragraph lengths	Short paragraphs, no established limits
Sentence lengths and structure	Short sentences, no established limits
Vocabulary considerations	Use the jargon associated with each department
Style Manual:	For graphics and source documentation, *The Style Manual of the American Psychological Association*, 4th ed., 1994
Specialized Dictionaries:	None
Specialized Encyclopedias:	None
Reviews Needed:	
Legal	Yes, by D. Davis, representing the City
Marketing	No
Engineering	No

Figure 3.3 A sample documentation plan. (*Continued*)

Research and Development	No
Training	No
Editing	Yes, by S. Jackson, representing the City
Consumer/Audience	No
Validation	Yes, by S. Jackson, S. Owens, & L. Mason
Usability testing:	None formally, but comments are provided by Municipal Board members throughout the year and will be incorporated into the next report.
Production:	
Type of production (e.g., saddle stitched binding, stapled pages, videotape, CD-ROM)	Photocopied pages, spiral bound, beige card stock cover
Time required	Two weeks, by in-house printing division
Cost (total and per unit)	$XX.00 per report x 53 reports = $XXX.00
Sources of Information: (Mark Primary and Secondary sources of information.)	
Personal observation	Previous experience writing this report
Personal experimentation	N/A
Interviews (and with whom)	Interviews with S. Jackson and D. Davis, who provide drafts of summaries written by each department
Questionnaires (respondents and sampling method)	No
Reference works	Previous reports for 1990-1992
Periodicals	N/A
Software	N/A
Models	N/A
In-house publications	As provided by departments, usually as a source of graphics for their summaries
Other:	

Figure 3.3 (*Continued*)

Documentation of These Sources:	Telephone and e-mail, documented in a research log by S. Owens or L. Mason
Timetable for Project:	
Information gathering	April 27, 1994
Organizing and compiling	May 1, 1994
Writing a first draft	May 10, 1994
Getting reviews	May 20, 1994
Writing a second draft	June 5, 1994
Getting reviews	June 15, 1994
Editing the draft	June 22, 1994
Validating the document	July 1, 1994
Usability testing the document	Ongoing, in-house
Getting post-validation reviews	July 5, 1994
Incorporating changes	July 10, 1994
Producing the document	July 15, 1994
Outline of the Project (tentative):	N/A
Project Distribution Date:	August 1, 1994

Figure 3.3 A sample documentation plan. (*Continued*)

This sample is a simple one, but it outlines all the responsibilities and checkpoints associated with one short printed document. Other projects require much more explanation and involve more team members. But the concept is the same. Develop a workable plan. Document everything. Refer to the doc plan often.

Summary

Some projects require less planning; for example, short memoranda can be planned entirely in your head, so you won't create a formal doc plan before you write a memorandum. But most projects are more extensive and may require you to plan hardcopy or softcopy projects involving one product or several complementary products. Especially as your company looks to documenting procedures and the resulting products associated with these procedures, you need to get in the habit of creating doc plans. If you know how to plan a document, whether you're managing the entire process or simply completing one small part of it, you'll always be aware of your responsibilities in making a project successful.

Copies of the doc plan should be distributed to all people involved with the project. By keeping your copy close by as you research the information and plan the form that information will take, you should eliminate detours and reach your destination on schedule.

Research Activities

1. Using the subject headings provided in this section, create a plan for a current project. Keep this plan for further reference.
2. Create an online form or guidelines for a doc plan so you can quickly create doc plans for future projects.

For Further Reading

Hackos, JoAnn T. 1993. *Managing Your Documentation Projects.* New York: John Wiley & Sons, Inc.

Reilly, Norman B. 1993. *Systems Engineering for Engineers and Managers.* New York: Van Nostrand Reinhold.

Schultz, Susan I. 1993. *The Digital Technical Documentation Handbook.* Burlington, MA: Digital Press.

CHAPTER FOUR

Ethics

Ethics has been and always should be an important part of any professional's vocabulary and approach to work, others—and life in general. It's become the focus of the popular media and professional world. Everyone seems interested in *ethics* and what is *ethical*. Especially in this decade, questions about values, legalities, and rights have become highly publicized. The lines among personal and public rights and responsibilities, individual and group rights, and accepted cultural codes of behavior have shifted or been smudged. We agree that having ethics is a good thing, and we analyze behavior and attitudes through variations on what is ethical (such as other professionals' previous emphasis on situational ethics). But who determines what is ethical? And how does ethics affect the research process, either on a daily, pragmatic basis or in longer term, theoretical experiments?

To take a purely rhetorical approach, we can check dictionaries to learn what others have defined as an acceptable usage of ethics. For example, ethics has been defined as "the rules or standards governing the members of a profession" in the *American Heritage Dictionary*, and the concept of *standards* is usually found in corporate policies and professional codes of conduct.

When many people think of ethics, however, they assume that morality or values play an important part. That's when ethics becomes problematic. Although some values may be generally accepted by a given society at a given time, not all members of a group, including a profession, have the same values. That does not necessarily mean that some professionals are unethical, act "immorally" according to other professionals' standards, or lack values—some professionals' values are simply different from others. Members of a profession usually agree in general about appropriate standards of conduct, but they vary greatly in their interpretation of these standards when they're applied to specific situations, because they may have different personal value systems. Ethics, then, becomes not only difficult to define, but difficult to standardize.

Add to that difficulty the fact that social values change. For example, cigarette smoking was an acceptable practice for many people, and popular media often portrayed the "good" characters as smoking. Even nonsmokers accepted others' smoking behaviors—until recently. A cultural shift is taking place in the United States, and society has shifted from accepting smoking to accepting smoking in limited locations to considering a ban on smoking. Social values and norms change within nations, regions, professions, and other cultural groupings, such as families, professional associations, corporations, and academic institutions. As individuals, we may not agree with various standards that we may be required to follow in our professional work. However, as long as those standards don't conflict with our personal ethics, we all feel more comfortable when we know that there are standards for professional behavior.

The concept of ethics may in general be defined as a code of conduct within a profession. In practice, technical communicators follow principles that guide their interaction with others and their approaches to working with the various kinds of information with which they are concerned. The purpose of this chapter is to provide you with some concepts and guidelines for developing your own ethical code of conduct as you pursue personal and professional research. In this chapter, we present ideas for developing methodical, documented research methods and building in safeguards for the ethical use of information and treatment of people.

Public Concern with Ethics in Research

Currently, public concerns about ethics extend beyond research, but research methods are increasingly scrutinized and made public. The public is interested in making corporations, institutions, and individuals accountable for their actions, including research. We live in an age when the public's demands for accountability equal demands for progress. The concern with ethics grows as the need for accountability is considered along with the need for immediate results. Researchers are frequently forced to face the reality as they see ethical results coming from ethical research procedures, but those who are awaiting the results see immediacy as the primary ethical concern.

Ethics in research is particularly problematic because research historically has been private, and the public generally hasn't demanded as much accountability from researchers as from other professionals. What occurred in the laboratory, for example, was shrouded in mystery until the researchers decided to publish the results of their experiments. Often secrecy was a necessity, as several independent researchers worked on a similar design or concept, and each feared another researcher would "borrow" methodology and thus become the first to solve a scientific or technological puzzle. Because technical or scientific research often has involved matters of national security, researchers were encouraged to keep their work hidden from the public, and possible spies or saboteurs.

The question of what should be researched has posed additional problems for technical or scientific researchers. Because resources are perceived as increasingly scarce, grant-funded research that comes in part from taxes is often the subject of media scrutiny. The public becomes concerned when grants of several hundred thousand dollars are awarded to researchers studying what many perceive as unimportant topics. When funds are tight, the public often questions the need for research that leads possibly to more knowledge, not to specific results. When children are hungry in this country, money allocated for studying linguistic patterns in e-mail messages, for example, may seem outrageous. How ethical is research when it leads only to more information, especially if that information does not immediately—or even eventually—improve the human condition? That question is often loudly and hotly debated.

Even when the need for research, such as medical research, is seen as important, and results that improve the quality of life as well as add to the knowledge base are anticipated, the appropriateness of the researcher's methods are often debated. Organizations like People for the Ethical Treatment of Animals (PETA), for example, demonstrate and actively campaign in the media to stop the use of animals in testing. Medical experiments, product testing, and other forms of research frequently use animals instead of humans or computerized simulations to test procedures and products. How ethical is the use of animals in testing?

Is such testing ethical if ultimately human lives will be saved or quality of life improved because of the results of these tests? Is it ethical—or just ethnocentric—to presume that humans are "worth" more than other species?

In many areas of technical and scientific research, ethical dilemmas are not easily solved. More often, the question about the propriety and necessity of research results in debates, which may lead to legislation, to change the nature of research.

Even in the daily research of many technical writers who primarily gather information from interviews or experience with products and processes, ethics can pose confusing questions. Someone may be misquoted, however innocently, in an in-house article, because the researcher/writer didn't have the time to double check the information. There may be discussions or disagreements among representatives from the legal, marketing, and technical communications departments about the need, for example, for hazard information in product literature. A marketing group may fear that bright danger notices may unfairly make consumers wary of a product, when the product is basically safe. Technical communicators may insist that some people in some situations may misuse the product and harm themselves. And the legal department may emphasize what is legally required, not what is perhaps the best way of protecting people or of promoting a product.

Another ethical issue is the use of information retrieved from a variety of sources. Technical communication provides many opportunities for plagiarism. For example, the use of boilerplate seems straightforward in daily technical communication. Technical writers, editors, and artists, among others, are encouraged to use boilerplate in corporate documents. In college-level writing, or other areas of professional writing, for that matter, copying information from one document and inserting it verbatim into another document would be called plagiarism and could be the basis of a lawsuit. Within a company that owns the copyright on the original material, however, using boilerplate is a convenience and is accepted practice. Nevertheless, individuals who use "boilerplate" from their own work, for example, in the multiple submission of articles for publication, find that this practice is usually frowned upon. Submitting the same information for publication in more than one place at a time is considered unprofessional, as is using sections of the same research document in multiple publications.

When we think about the possibilities of uploading, downloading, and sharing information through networks, we increase the confusion of what is acceptable to use because it's public domain or "shareware" and what is copyrighted or copyrightable. When is it okay to take information from an e-mail message and use it in your publication, when it was someone else's expression of an idea? When is it okay to download graphics or text files and use them in your own work? And is it okay when you use someone else's ideas or information but don't profit from them? Or is it always unethical to take information and attribute it to yourself, even if it came from other, often unnamed and difficult to attribute sources? The proliferation of electronic media and the amounts of information to which technical communicators have access only makes our concerns about ethics more important—and more confusing.

One common trend in research is absolute accountability and greater access to information. Researchers are accountable for their methods to the agencies and institutions for which they conduct research. However, they also have a responsibility to help explain to the public the nature of their research and the safeguards they've built into their methodology. Getting permission for using information or following a research method, as well as documenting research as it takes place, are standard practices in the lab and in the workplace.

Codes of Conduct

Many professional societies, among them the Society for Technical Communication, emphasize ethics and ethical behavior in their publications and committees. The Society for Technical Communication's Code for Communicators (Figure 4.1) provides guidelines for its members' professional conduct. Members of the Society, for example, mutually agree that these guidelines are good, but the Society doesn't track down someone who has "violated" these principles. Guidelines like the Society for Technical Communication's and corporations' codes of conduct are merely frameworks for ethics; ethics itself is the way we interpret these codes through our conduct and thought.

Another well-known professional association, the American Psychological Association (APA), also promotes a code of ethics to guide its members in their research. A statement of the "Ethical Principles of Psychologists and Code of Conduct" was published by the APA in December 1992 in the journal, *American Psychologist.* In the most recent edition of *The Publication Manual of the American Psychological Association,* relevant sections of these ethical standards are reprinted, so that researchers using the

STC society for technical communication

Code for Communicators

As a technical communicator, I am the bridge between those who create ideas and those who use them. Because I recognize that the quality of my services directly affects how well ideas are understood, I am committed to excellence in performance and the highest standards of ethical behavior.

I value the worth of the ideas I am transmitting and the cost of developing and communicating those ideas. I also value the time and effort spent by those who read or see or hear my communication.

I therefore recognize my responsibility to communicate technical information truthfully, clearly, and economically.

My commitment to professional excellence and ethical behavior mans that I will

- Use language and visuals with precision.
- Prefer simple, direct expression of ideas.
- Satisfy the audience's need for information, not my own need for self-expression.
- Hold myself responsible for how well my audience understands my message.
- Respect the work of colleagues, knowing that a communication problem may have more than one solution.
- Strive continually to improve my professional competence.
- Promote a climate that encourages the exercise of professional judgment and that attracts talented individuals to careers in technical communications.

Figure 4.1 Code for communicators.

manual are made aware, if they weren't already, of the standards members of their profession are expected to uphold. As the authors of the manual state (p. 292), "basic ethical principles underlie all scholarly writing. These long-standing ethical principles are designed to achieve two goals: 1. To ensure the accuracy of scientific and scholarly knowledge and 2. To protect intellectual property rights." The standards outlined in the APA's code concern the ways to report research results, avoid plagiarism, give publication credit to other researchers, report data in more than one publication simultaneously, make results available to other professionals for verification, maintain confidentiality during the review process when the results of new research are being considered for publication, and work safely and professionally with people and animals involved in research.

Codes of conduct, whether promoted by professional associations or corporations, indicate the need for a standard of behavior that members of a specific profession should uphold. In principle, most professionals agree to the statements of honesty, integrity, and mutual cooperation emphasized in these codes. But in daily practice, the codes of conduct may be too general to apply to specific, tricky ethical dilemmas. You need to have your own code of personal ethics to help you determine what you will and will not do on the job and in your working relationships.

Strategies for Developing Your Code of Ethics

The following four strategies can help you work ethically and methodically as you conduct qualitative or quantitative research, whether for the creation of documents or products or for experimentation leading to more knowledge.

1. Get approval for your research methodology.
2. Methodically and accurately conduct the research.
3. Open your information to other experts.
4. Give credit where needed and/or appropriate.

Get Approval for Your Research Methodology

Approval may take many different forms, depending if you're conducting day-to-day research that takes place within your company, freelancing information that will turn into a document, or planning experimental research. But you still need to get some form of approval, and if you can document your method, so much the better.

Approval within a Company

When you're working within a company, usually the employees' handbook or policy manual will provide the guidelines for getting approval. Procedures are usually documented and in place for all the types of information gathering and checking necessary prior to final publication or presentation.

If your work as a technical writer requires you to gather information primarily from in-house sources, you probably have a standard method you follow to get information and check it. The process of approving information before it's released in final form may be a team project, involving technical writers, subject matter experts (SMEs), technical editors, and possibly the legal department. These kinds of reviews are significant because they provide an opportunity for numerous people with different perspectives to check the information for accurate, honest content and presentation. But they're also excellent opportunities for ensuring that the information is designed, written, and presented in a form that's accessible to the audiences for whom it was intended.

Technical writers usually have their own processes for checking the information they gather before they put it into a draft. They may talk with SMEs, for example, and then ask the SMEs about any information they question or to get updated information since the interview. SMEs may be required, according to corporate policy, to sign off on a draft, so that the information has been guaranteed as accurate. Technical editors may also be responsible for double checking information. And the legal department may routinely check information designed for out-of-company distribution, to make sure that the information is appropriate and that all questions about liability have been resolved.

Within a company, approval may take the form of signature lines and sign-offs on drafted documents. If approval is needed from outside sources, such as other companies, the legal department may be directly involved with approval for copyrighted, trademarked, and patented information or materials. Release forms signed by persons outside the company who were interviewed or who submitted information for publication or presentation may be kept on file.

Approval for Freelancers

When you work as a freelance technical writer or editor, you usually follow someone else's (such as the client's or publisher's) procedures for checking and getting approval of information. You may be provided with forms by which you secure copyright permission, for example. If you're writing a document for a publisher, and you plan to use information that has been published elsewhere (such as a copy of a document or quotations from experts), the publishing company provides you with standard forms for securing copyright permission. The person or company providing the permission may require you to pay a fee for the use of the information (whether text or graphics), state the permission in a specific style or format, or both.

If you submit information for publication, such as a freelance article submitted to a journal for publication, you may be required to sign forms stating that the information is original and has not been submitted for publication elsewhere. You are legally liable for the accuracy of your information, and the journal's publishing agency requires you to sign an agreement that the information is yours. It is your responsibility to ensure that you check the information, so you probably want to document the research you completed to write the article. Often, you also are required to sign a copyright release form for publishers, including professional associations that publish journals and conference proceedings. The copyright then is owned by the publisher, and you also have to get permission to use the information in your article or paper if you want to quote it in another publication or make several copies for distribution.

Approval for Experimental Research

The federal government has set requirements for approving research involving people, and universities, colleges, and research institutions follow these requirements when they develop their approval guidelines and forms. Generally, you are required to develop and have approved methods that ensure the safety of the people participating in their research. If your methods change, you have to get approval for a change in procedure. You also have to make sure that your potential participants are aware of the type of research you're conducting. Participants thus have to give *informed consent* that they agree to participate in your research project.

When you request approval of your research, you submit the appropriate forms and may have to talk directly with the committee or board that approves the project. At a university or college, the review board usually is made up of faculty and administrators who are very familiar with research methods and federal regulations. They typically represent different disciplines and can provide a broad perspective about the appropriateness of the research. If you're getting approval for your research from a granting agency, for example, usually your methods are approved by a panel of experts who are familiar with the subject and the appropriate methodologies for research within that discipline.

Documents like Figure 4.2 outline your responsibilities as the researcher, the procedures for getting approval for your project, the acceptable parameters for research methods, and the requirements for getting informed consent from participants. When you're applying for formal approval from a human subjects review board or other approval-granting committee, you first request the published guidelines, like these, before you design the research. Then you propose your project and answer any questions you might be asked about your research design.

The checklist in Figure 4.3 summarizes some of your responsibilities as a researcher seeking approval for your methods and the information you gather.

Methodically and Accurately Conduct the Research

As you design research, you should always make sure that the participants and the equipment or materials you're using are as safe as possible. Before you begin to ask for approval of your methodology, you need to assure yourself that your methods are as risk free as they can be and that they are sound. Establishing guidelines and procedures for your research, making sure that your approach is logical and methodical, and considering the safety of any participants in your research are essential parts of the project-design process. Then you apply for approval from the agencies or committees that need to give a formal okay before you begin the project.

A part of ethics can also be good sense. When your research has been planned and approved, you and any others involved with the project should stick with the plan. If you are conducting interviews with persons selected as participants and more than one person will be interviewing participants, you make sure that all interviewers use the same style,

Office of Sponsored Programs and Research
106 University Hall
Bowling Green, Ohio 43403-0183
Phone: (419) 372-2481
FAX: (419) 372-0304
E-mail: spar@bgnet.bgsu.edu

HUMAN SUBJECTS REVIEW BOARD
GUIDELINES FOR THE REVIEW OF RESEARCH, DEVELOPMENT AND RELATED ACTIVITIES

I. PURPOSE/DEFINITION OF HUMAN SUBJECTS

When people are involved as subjects in research or related activities (see Definition section VIII) conducted under University auspices, both the institution and the individual project director incur legal and ethical responsibility for assuring that the rights and welfare of participants are adequately protected. This responsibility extends to any mode of systematic investigation, whether sponsored solely by the University or funded externally, and conducted either on or off campus. Federal law defines a human subject as "a living individual about whom an investigator (whether professional or student) conducting research obtains (1) data through intervention or interaction with the individual, or (2) identifiable private information." (45 CFR 46.102 (f).) "Intervention" includes both physical procedures and manipulations of the subject or the subject's environment. "Interaction" includes any kind of communication or interpersonal contact between investigator and subject.

II. HUMAN SUBJECTS REVIEW BOARD (HSRB)

In compliance with federal regulations the University has established an institutional Human Subjects Review Board (HSRB), administered through the Office of the Associate Vice President for Research, to oversee its obligations with respect to human subjects. The jurisdiction of the HSRB includes all research and related activities covered under these guidelines. No project involving human subjects should proceed without the explicit written approval of this Board. This approval is obtained by completing and submitting for review a written protocol describing a) the research purposes and methods; b) the selection and characteristics of the subjects; c) procedures for gaining informed consent; d) procedures for protecting confidential- ity of information collected; and e) potential risks to the subjects and procedures for minimizing those risks. The Board reviews protocols on a monthly basis.

III. TYPES OF ACTION ON PROPOSED PROJECTS

In acting upon proposals, the HSRB may take any one of five actions.

1. Exempt the research from full review.

Federal regulations provide that certain kinds of research (e.g., some surveys, field interviews, observations, evaluations of standard educational practices, or tests) involving no more than minimal risk to subjects can be exempt from full HSRB review and record keeping (see Section IX). Final determination of exempt status is the responsibility of the Review Board. Therefore, all investigators need to complete the attached forms in order to provide the information necessary to determine if exemption is applicable. In no case, however, is research involving children, prisoners, pregnant women, or handicapped persons eligible for exemption.

2. Approve the research procedures.
3. Approve the research procedures subject to modifications.
4. Disapprove the research procedures.
5. Defer action on the proposal pending receipt of additional information or further clarification of specific items as may be identified.

For previously approved projects, any additions or changes in human subject involvement or procedures must be brought to the attention of the HSRB, which will then consider whether or not the original assessment should be modified in any way. If so, the HSRB will treat the matter as a new case and no new procedures should be implemented until Board approval has been obtained.

Figure 4.2 Human subjects research board guidelines.

-2-

Each project not granted exempt status will be reviewed annually by the HSRB to assure that investigators continue to follow approved procedures and practices regarding the use of human subjects. If subjects behave as if personal rights have been violated in any form, the investigator should inform the HSRB.

If a proposal or procedure receives final disapproval by the HSRB and the principal investigator wishes a further hearing on the matter, an appeal may be made to the Associate Vice President for Research. In this case an ad hoc appeals committee will be convened, consisting of the Chair of the HSRB, and two or more persons who have previously served on the HSRB as well as any special consultants that may be required. The HSRB will also hear complaints from any subjects who feel their rights have been violated.

IV. TYPES OF REVIEW FOR PROPOSED PROJECTS

A. Exempt Projects

Research involving any of the modes listed in Section IX must be submitted to the HSRB for approval as exempt. Therefore, partial completion of the HSRB protocol form is required and investigators are encouraged to pay close attention to the directions accompanying that form.

B. Expedited Review

All research not qualifying for an exemption must be reviewed by the HSRB and all sections of the form must be completed. To expedite the review process, members and alternates to the HSRB have been designated as area reviewers for pre-screening of proposals. If at least 2 reviewers determine that no more than minimal risk is involved, and the research falls into one of the expedited categories (see section X), tentative approval may be given and the project may proceed subject to final HSRB approval. However, the HSRB will review all projects in a convened meeting, and the final evaluation of each proposal will be made by this body, not by the area representative. For further information about types of research qualifying for expedited review, see section X or contact your area representative.

C. Full Board Review

Any research not exempt or expedited must obtain Full Board review and approval. Therefore, no research activities involving human subjects on such projects may proceed until Full Board approval is obtained. All projects seeking external financial support (not including Thesis/Dissertation Research Support) are defined by BGSU as requiring Full Board review and approval. 22 copies of the completed application must be submitted to the Research Services Office. Investigators are urged to meet this requirement prior to submission of external proposals because many agencies have not adopted, or do not plan to use the 60 day grace period of HSRB certification used by the Department of Health and Human Services. Many agencies require HSRB certification at the time the proposal is submitted.

Copies of the application for approval of research involving human subjects will be distributed to each HSRB member for review prior to the next convened meeting. The investigator may be asked to answer questions or meet with the HSRB to clarify specific points regarding the involvement of human subjects in the proposed activity. Conversely, if specific questions are anticipated as a consequence of the proposal, the investigator may request a meeting with the HSRB prior to the typing of a final draft. This meeting will not necessarily take the place of a regular review.

V. CRITERIA FOR HSRB APPROVAL

In reviewing activities covered under these guidelines, the HSRB seeks to determine that all of the following requirements are met.

1. Risks to subjects are minimized and reasonable.

   Research procedures should be consistent with sound research design, should not expose the subjects to unnecessary risk, and when possible, should be the same as

Figure 4.2 Human subjects research board guidelines. (*Continued*)

-3-

those already being performed on the subjects for diagnostic purposes.

The benefits to the subjects are considered as well as the importance of the knowledge that reasonably may be expected to result.

2. Selection of subjects is equitable.

   The purposes of the research and the setting in which the research is conducted are considered. Indications of coercion or prejudice must be absent, and participation must be clearly voluntary.

3. Informed consent is sought and documented from each prospective subject or from the subject's legally authorized representative.

4. Provision is made for collecting, utilizing, and storing data in a manner that protects the safety and privacy of the subjects and the confidentiality of the data.

5. Appropriate safeguards are included to protect the rights and welfare of the subjects.

   This last point applies particularly to activities where subjects are likely to be vulnerable to coercion or undue influence, to projects involving children, subjects with mental or physical illness, or involving persons who are economically or educationally disadvantaged.

Once the HSRB determines that all of the above points have been satisfied, approval may be given for the project to proceed subject to further review or disapproval by other University officials. However, other officials may not approve the project if it has been disapproved by the HSRB.

The HSRB may suspend or terminate approval of research that is not being conducted in accord with HSRB requirements or that has been associated with unexpected harm to subjects. Any suspension or termination shall include a statement of reasons for such action and shall be reported promptly to the investigator, to appropriate university officials, and to the funding agency, if applicable.

VI. RESPONSIBILITY OF THE INVESTIGATOR

The investigator must secure the necessary HSRB approval prior to implementing a proposed project. This is accomplished by completing the application for approval of research involving human subjects and submitting this to a member of the HSRB or directly to the Research Services Office. (One copy of the completed Summary should be retained by the investigator.) Under no circumstances will the University endorse participation by subjects in research or other projects until the application has been submitted and approved by the HSRB. Furthermore, if the principal investigator desires to introduce procedures not assessed previously or to change the approved procedures in such a way as to raise anew questions of the treatment of human subjects, such additions or changes must be brought to the attention of the HSRB Chair.

VII. INFORMED CONSENT OF RESEARCH SUBJECTS

A. Obtaining and Documenting Consent

Unless a waiver has been approved by the HSRB, the investigator must obtain and document in writing the subject's informed consent. The key elements of "informed consent" are:

1. A statement that the study involves research, an explanation of the purpose of the research and the expected duration of the subject's participation, a description of the procedures to be followed, and identification of any procedures which are experimental;

2. A description of any reasonably foreseeable risks or discomforts to the subject;

Figure 4.2 (*Continued*)

-4-

3. A description of any benefits to the subject or to others which may reasonably be expected from the research;

4. A disclosure of appropriate alternative procedures or courses of treatment, if any, that might be advantageous to the subject;

5. A statement describing the extent, if any, to which confidentiality of records identifying the subject will be maintained;

6. For research involving more than minimal risk, an explanation as to whether any compensation and an explanation as to whether any medical treatments are available if injury occurs and, if so, what they consist of, or where further information may be obtained;

7. The name of the person to contact for answers to pertinent questions about the project and the subject's rights, and whom to contact in the event of a research-related injury;

8. A statement that participation is voluntary, refusal to participate will involve no penalty or loss of benefits to which the subject is otherwise entitled, and the subject may discontinue participation at any time without penalty or loss of benefits to which the subject is otherwise entitled.

For research involving minors - i.e. individuals who have not attained the legal age of consent (18) - a parent or guardian must be given a pre-service briefing in written or oral form, and a consent form (See Appendix A) must be obtained for each child. In the case of children over seven years old, the child must also give assent. In research where the minor is old enough to give fully informed consent to non-sensitive, non-risky research procedures, researchers may request waiver of parental consent. Waiver of parental consent can only be granted by the HSRB.

Because the forms used as documentary evidence of informed consent must be preserved by the principal investigator at least 36 months following termination of the project, according to Federal Guidelines, such forms are required to be written in the affirmative. Forms which provide only for denial or refusal of participation are unsatisfactory evidence of informed consent by participating subjects. Therefore, the HSRB requires the use of consent forms which document a subject's affirmative decision to participate in research activities, or a parent's or guardian's affirmative permission for a child's participation in the research activity.

The investigator should specify to the HSRB the procedures which will be used to preserve the confidentiality of these records. If the investigator leaves campus before the 36 months have expired, copies of the appropriate documents should be filed with the Graduate College where they will be secured in locked files. If the project was carried out at another institution, the files may be retained at that institution, but the HSRB should be informed of their location. In any case, records are subject to audit by University as well as federal officials and should be retained for easy and quick access if requested.

Some research may be so indirect, innocuous and innocent of imposition on the rights and welfare of human subjects as to make informed consent a moot requirement. Therefore, the HSRB may choose to waive this requirement. However, such action must be based upon clearly defensible grounds, unrelated to expediency of the research, and the principal investigator must include these justifications in the application submitted to the HSRB. These criteria also apply to waiver of parental consent for research with minors, as well.

B. Special Conditions of Informed Consent

1. Research Involving Deception: In some research it is not possible to fully inform the subjects of the experimental procedures without destroying the validity of the research. In such cases informed consent need not be based on full pre-service information. The HSRB will set limits to the incompleteness of such information.

The principal investigator or a fully informed assistant must give each subject an explanation to questions arising after the experiment. This should occur immediately following such service for each subject. However, if in the judgment of the HSRB such information could adversely affect subsequent data collection in the same study, the full explanation may be delayed for a reasonable period of time.

Figure 4.2 Human subjects research board guidelines. (*Continued*)

-5-

In those cases where it is necessary to seriously mislead the subject and/or where it is possible that the experimental treatment may result in emotional stress for the subject, a full debriefing is mandatory immediately following service. In the case of minors, in addition to the individual explanation, an explanation should be provided each parent.

2. Research Involving Stressful Stimuli: In those studies which use stressful stimuli, the HSRB must be assured that the duration and intensity of such stimuli are within acceptable limits.

In those studies which involve emotional stress, the HSRB must be assured that adequate safeguards are provided in the research procedures to control the severity of such stress reactions and that an adequate debriefing procedure will be used immediately after the conclusion of each subject's service.

3. Research Involving Minors: Special protections are afforded minors (persons under age 18), including the need to obtain parental consent. If the research is not sensitive and little or no risks are involved, waiver of parental consent may be requested by the investigator. For parental consent to be waived, the HSRB must receive a written request and formally approve it.

VIII. DEFINITION OF TERMS

RESEARCH means a systematic investigation, including research development, testing and evaluation, designed to develop or contribute to generalizable knowledge. Activities which meet this definition constitute research for purposes of this policy, whether or not they are conducted or supported under a program which is considered research for other purposes. For example, some demonstration and service programs may include research activities.

HUMAN SUBJECT means a living individual about whom an investigator, whether professional or student, conducting research obtains (1) data through intervention or interaction with the individual, or (2) identifiable private information. "Intervention" includes both physical procedures and manipulations of the subject or the subject's environment. "Interaction" includes any kind of communication or interpersonal contact between investigator and subject.

AT RISK A subject is considered to be at risk if the possibility of physical, psychological, sociological, or other types of harm may be the consequence of an activity which goes beyond the application of established and accepted methods necessary to meet the needs of the subject, or which increases the ordinary risks of daily life, including the recognized risks inherent in a chosen occupation or field of service.

MINIMAL RISK means that the probability and magnitude of harm or discomfort anticipated in the research are not greater in and of themselves than those ordinarily encountered in daily life or during the performance of routine physical or psychological examinations or tests.

INFORMED CONSENT is the agreement, usually obtained in writing from a subject, or an authorized representative, to the subject's participation in an activity. Such consent can be "Informed" only under circumstances that provide the subject or representative sufficient opportunity to consider whether or not to participate and that minimize the possibility of coercion or undue influence.

IX. EXEMPTIONS

On June 18, 1991, regulations amending basic Health and Human Services policy for the protection of human research subjects were published in the Federal Register (56 No. 117). These regulations contain exemptions for broad categories of research which involve little or no risk to research subjects (see listing below). Responsibility for approving exemptions rests with the Review Board, but initial determinations are made by the investigator in cooperation with the area representative. **In no case, however, is research involving children, prisoners, pregnant women, or handicapped persons eligible for exemption.**

Unless the research is covered by other subparts of the federal regulations, exemption from general Human Subjects requirements is possible for research activities in which the only involvement of human subjects will be in one or more of the following categories:

Figure 4.2 (*Continued*)

-6-

{1} Research conducted in established or commonly accepted educational settings and involving normal educational practices, such as (i) research on regular and special education instructional strategies, or (ii) research on the effectiveness of or the comparison among instructional techniques, curricula, or classroom management methods.

{2} Research involving the use of educational tests (cognitive, diagnostic, aptitude, achievement), survey or interview procedures, or observation of public behavior, unless: (i) Information obtained is recorded in such a manner that the subjects can be identified, directly or through identifiers linked to the subjects; and (ii) any disclosure of the subjects' responses outside the research could reasonably place the subjects at risk of criminal or civil liability or be damaging to the subjects' financial standing, employability, or reputation.

{3} Research involving the use of educational tests (cognitive, diagnostic, aptitude, achievement), survey or interview procedures, or observation of public behavior that is not exempt under paragraph {2} if: (i) The subjects are elected or appointed public officials or candidates for public office; or (ii) federal statute(s) require(s) without exception that the confidentiality of the personally identifiable information will be maintained throughout the research and thereafter.

{4} Research involving the collection or study of existing data, documents, records, pathological specimens, or diagnostic specimens, if these sources are publicly available or if the information is recorded by the investigator in such a manner that subjects cannot be identified, directly or through identifiers linked to the subjects.

{5} Research and demonstration projects which are conducted by or subject to the approval of federal department or agency heads, and which are designed to study, evaluate, or otherwise examine: (i) Public benefit or service programs; (ii) procedures for obtaining benefits or services under those programs; (iii) possible changes in methods or levels of payment for benefits or services under those programs.

{6} Taste and food quality evaluation and consumer acceptance studies, (i) if wholesome foods without additives are consumed or (ii) if a food is consumed that contains a food ingredient at or below the level and for a use found to be safe, or agricultural chemical or environmental contaminant at or below the level found to be safe, by the FDA or approved by the EPA or the Food Safety and Inspection Service of the U.S. Department of Agriculture.

X. EXPEDITED REVIEW

Certain research activities involving no more than minimal risk and in which human subjects are involved in special ways are approved by Federal guidelines (46 FR 8392) for "expedited review." The categories acceptable for expedited review are listed below:

{1} Collection of: hair and nail clippings, in a non-disfiguring manner; deciduous teeth; and permanent teeth if patient care indicated a need for extraction.

{2} Collection of excretal and external secretions including sweat, uncannulated saliva, placenta removed at delivery, and amniotic fluid at the time of rupture of the membrane prior to or during labor.

{3} Recording of data from subjects 18 years of age or older using non-invasive procedures routinely employed in clinical practice. This includes the use of physical sensors that are applied either to the surface of the body or at a distance and do not involve input of matter or significant amounts of energy into the subject or an invasion of the subject's privacy. It also includes such procedures as weighing, testing sensory acuity, electrocardiography, electroencephalography, thermography, detection of naturally occurring radioactivity, diagnostic echography, and electroretinography. It does not include exposure to electromagnetic radiation outside the visible range (for example, x-rays, microwaves).

{4} Collection of blood samples by venipuncture, in amounts not exceeding 450 milliliters in an eight-week period and no more often then two times per week, from subjects 18 years of age or older and who are in good health and not pregnant.

Figure 4.2 Human subjects research board guidelines. (*Continued*)

-7-

{5} Collection of both supra- and sub-gingival dental plaque and calculus, provided the procedure is not more invasive than routine prophylactic scaling of the teeth and the process is accomplished in accordance with accepted prophylactic techniques.

{6} Voice recordings made for research purposes such as investigations of speech defects.

{7} Moderate exercise by healthy volunteers.

{8} The study of existing data, documents, records, pathological specimens, or diagnostic specimens.

{9} Research on individual or group behavior or characteristics of individuals, such as studies of perception, cognition, game theory, or test development, where the investigator does not manipulate subjects' behavior and the research will not involve stress to subjects.

{10} Research on drugs or devices for which an investigational new drug exemption or an investigational device exemption is not required.

XI. REFERENCES

1. "Basic HHS Policy for Protection of Human Subjects," Federal Register 46, No. 16, January 26, 1981 (46 FR 8366), and March 8, 1983 (48 FR 9269).
2. The Institutional Guide to DHEW Policy on Protection of Human Subjects, Department of Health, Education, and Welfare, Publication 72-102, December 1, 1971. (No. 596)
3. Ethical Principles in the Conduct of Research with Human Participants. Washington, D.C.: American Psychological Association, 1973.
4. "Ethical Issues in Research with Human Subjects." Science 155, 47 (1967).
5. "Privacy and Behavioral Research." American Psychologist 21, 437 (1966).
6. "Declaration of Helsinki." Journal of the American Medical Association 12, 32 (1966).
7. "Ethical Standards of Psychologists." American Psychologist 18, 56-60 (1963).
8. "Protection of Human Subjects." Federal Register 56, No. 117, June 18, 1991.

XII. SAMPLE CONSENT INFORMATION FORM: This form is intended only as a guide. Researchers should present the required information in the most appropriate format to the items enclosed in parentheses. Both the participant and researcher should retain a copy of the signed consent form. Note that research involving minors requires written consent from the parent/guardian and from the minor if the child is over seven years of age.

I consent to serve as a subject in the research investigation entitled: (title). The nature and general purpose of the study have been explained and the attached statement has been read to me by (name of researcher), from (department).

I understand the purpose of this research is (give brief explanation), and that the research procedures involve (duration and experimental procedures).

The potential benefits and risks to participants in this project are: (give brief explanation).

I understand that my participation is voluntary and that all information is confidential and my identity (or the identity of my child) will not be revealed; I (or my child) am/is free to withdraw consent and to discontinue participation in the project at any time; any questions I/my child may have about the project will be answered by the researcher named below or by an authorized representative.

Bowling Green State University and the investigator named below have responsibility for ensuring that participants in research projects conducted under university auspices are safeguarded from injury or harm resulting from such participation. If appropriate, the person named below may be contacted for remedy or assistance for any possible consequences from such activities.

On the basis of the above statements, I/my child agree(s) to participate in this project.

______________________ Participant's Signature	______________________ Researcher's Signature
	Address
______________________	Home Telephone / Campus Telephone
Date	

Figure 4.2 ***(Continued)***

-8-

XIII. **COMMITTEE MEMBERSHIP**

Feel free to contact any of the following HSRB committee members for assistance:

Dr. Anthony Boccanfuso	1995	Director, Research Services	372-2481
Mr. Curtis Brant	1995	Psychology	372-2301
Dr. Julie Burke	1996	Interpersonal Communication	372-2406
Dr. Rich Clark	1995	Criminal Justice	372-2326
Dr. Michael Coomes	1997	Higher Ed & Student Affairs	372-7157
Dr. Arch Darrow	1995	Management	372-2986
Ms. Nancy Footer	1995	Legal Affairs	372-0464
Dr. Laura Juarez deku	1997	Biology	372-8559
Dr. Josh Kaplan	1995	Student Health Services	372-2277
Dr. Jennifer Kinney, Chair	1995	Gerontology	372-7768
Mr. Kip Miller	1995	Wood County Hospital	352-6571
Dr. Clifford Mynatt	1996	Psychology	372-8309
Mr. Raymond Nagy	1995	Prisoner's Advocate Representative	352-9170
Ms. Joanne Navin	1996	Student Health Services	372-7429
Dr. William O'Brien	1995	Psychology	372-2974
Dr. Milagros Pena	1997	Sociology	372-2415
Dr. Melissa Spirek	1996	Telecommunications	372-8641
Dr. Phil Titus	1995	Marketing	372-8406
Dr. Ralph Wahrman	1995	Sociology	372-7262
Dr. Richard Wilson	1995	EDSE	372-7276
Dr. Jong Yoon	1997	Biology	372-2742

XIV. **WHAT TO SUBMIT**

The original signed application (with appropriate attachements) is due in Research Services, 120 McFall Center, by the last business day of the month prior to the month you anticipate the start of data collection. For those studies that anticipate or are receiving external funding, the original and 22 copies of the completed application are to be submitted.

Figure 4.2 Human subjects research board guidelines. (*Continued*)

Within a Company

- ☐ Read the company's policy manual or other policy statements to learn more about their practices and any published ethical codes.
- ☐ Follow an established, or develop and document, a standard procedure for gathering information, recording information, and double checking the information before it is presented or published.
- ☐ Check with the appropriate departments or groups, such as the legal department, to ensure that the documentation styles and formats, as well as proper procedures, are being followed.
- ☐ Give credit to the people whose ideas you've used in a presentation or a publication.
- ☐ Have the person who provided the information sign off on it to ensure its accuracy before you publish or present it.

If You're a Freelance Technical Communicator

- ☐ Ask to read the policy manual, handbook, or ethical guidelines developed by the company or client for whom you're working.
- ☐ Follow any established procedures for gathering information, recording information, documenting sources, and inserting special notices, such as copyright information.
- ☐ Develop your own standard procedures for gathering information, recording information, documenting sources, and developing special notices, such as copyright or patent information. Provide copies of your standard operating procedures to clients.
- ☐ Give credit to the people who provide you with information.
- ☐ Double check with the source all information you plan to use in a presentation or publication.
- ☐ Submit your material to only one publisher at a time. Wait to learn if that publisher wants the material before you offer it to another publisher.
- ☐ Follow the copyright guidelines about copying and reusing your materials after they have been published.
- ☐ Keep your research for at least a year after you've written or submitted material.

Figure 4.3 Checklist for conducting research ethically.

For Experimental Research

- ❑ Develop a safe, methodical approach to the experiment.
- ❑ Gain approval for your research by presenting your research plan to the appropriate review board.
- ❑ Document every part of the research process.
- ❑ Gain informed consent from any person participating in the experiment.
- ❑ Follow exactly the research plan that has been established and approved.
- ❑ Consult with the appropriate experts to interpret the results of your experiment.
- ❑ Make your results available to other researchers for verification.
- ❑ Give credit to others who provided you with information or ideas for your experiment.
- ❑ Accurately report your results.
- ❑ If you submit your findings for publication, submit your work to only one publisher at a time. Wait to learn if that publisher wants the material before you offer it to another publisher.
- ❑ Follow the copyright guidelines about copying and reusing your materials after they have been published.
- ❑ Keep your research for at least a year after you've written or submitted material.

Figure 4.3 Checklist for conducting research ethically. (*Continued*)

follow the approved procedure, and record information the same way. To ensure consistency in conducting interviews, for example, you may hold training sessions so that all interviewers know what to do and you can check the accuracy of each interviewer's method in accordance with the approved method. If your research requires raters to evaluate a product or a performance, you may want to conduct tests to ensure interrater reliability, have methods for checking raters' work, or hold a training session. As much as possible, you want to make sure that you and anyone else conducting the research follows the approved methods and does so reliably.

You also want to use valid, reliable instruments in your research. If you're using a questionnaire to gather information, you first test the questionnaire to make sure that it's a reliable instrument for gathering the information. When you've gathered the information and run statistical analyses of the results, you want to make sure you're using the appropriate type of analysis for the kind of research you've conducted. If you're testing a product, for example, and you have to take measurements of dimensions, speed, temperature, or other elements of the product, check the calibration of the measuring tools to ensure their accuracy. No matter what your "tools" are, if you're not sure that they are accurate and appropriate, you should consult with experts who can direct you to the appropriate instruments.

When you conduct research, you may have an idea of the type of results you may get, but research should be designed to test hypotheses, not to generate the results you anticipate. Not only is that not ethical, but it isn't a smart research practice. Research is a process of gathering information. Sometimes the results are encouraging in that they support your hypothesis; at other times, they may not support your original ideas. An important part of ethically conducting research is dealing with the results you get, no matter what they are.

Reporting information accurately is also a requirement of good research. When you have results, you want to evaluate them accurately and present them fairly. That means you present the results as you received them. Don't play down some findings that may be important, even if they were unexpected or they differed from your previous findings or ideas about what you'd find. Don't play up a minor point and emphasize it instead of other findings. Sometimes your findings may be just what you were hoping to learn, but you know that your work may not be generalizable to other situations. When you present your findings, you should be specific about the results, their application, and the possible flaws in or limitations of your research design. In short, you present your findings fairly.

Open Your Information to Others' Scrutiny

One reason for conducting research—of any type—is to share your findings. Good researchers make their information available to other experts who can learn from it, try to replicate it, criticize it, and build upon it. For example, technical communicators who want to share information about their communication processes and what's worked for them on the job may present this information at a conference. Their colleagues then have the opportunity to ask questions, make comments, and share their experiences. The conference becomes an opportunity for learning and sharing, based on information that once pertained to technical writing in one company.

Researchers who want to report the results from questionnaires they developed and distributed may publish their findings in a journal. Colleagues who read the journal then may write letters to the editor or discuss the article directly with the author. Especially through letters to the editor, colleagues criticize, praise, or provide different perspectives on the recent research. This dialogue may promote additional or new research in the subject.

Courtesy may not be a formal part of professional ethics, but it should be part of your interaction with others. As with all professional discussions, questioning someone's research or debating an issue should take place within the bounds of professional courtesy. Calling attention to research methods, findings, interpretations, and styles of presentation provides opportunities for professionals to discuss issues. Directing arguments against individuals personally in a public forum, however, is unprofessional. Discussions focused on the research, not the researcher, are more productive to the individual researcher and other professionals.

Technical communicators working on a grant-funded project may submit periodic progress reports to the granting agency, which may make the information available to the public or to other appropriate groups and agencies. Government-funded research projects may be audited by outside experts, who then report their evaluation of the research and make recommendations about the future of the specific project or the type of research. So at many levels, you may be required to or may simply choose to take the opportunity to open your research to others' scrutiny.

That's another good reason for keeping detailed records of the research in progress. Your documentation provides a trail that others can follow, not only to evaluate what you've done, but possibly to learn from it and replicate your research. Especially in technical or scientific experiments, replication is a good way to support a hypothesis or test a theory. Additional evidence adds to the development of knowledge about a subject, and your work can only be replicated if you've provided detailed information about your methods and materials that others can follow.

An ethical part of research behavior, then, is being able to take criticism. In fact, you should look for criticism. Worthwhile ideas stand up to criticism from people with different perspectives and experiences. The more a concept is tested and holds up to the tests, the better that concept is accepted. Your job as a researcher is to make sure that you've done the best you can do to keep records of your research and sources and to invite experts to evaluate your information.

Give Credit to Others

Good ideas don't exist in a vacuum, and everything new is based on something that went before. When you write a new manual, for example, your work is based on the information provided by several SMEs, your experience with similar products, and your prior experience writing manuals. But you didn't get that experience and knowledge all on your own. Throughout your life, you've read other manuals, studied manual writing, talked with people who've had good—and bad—experiences following directions from other manuals, and heard the jokes about "when all else fails . . . ." Being an ethical researcher means that you give credit where it's due.

Credit can take many different forms. It may simply mean thanking SMEs informally the next time you see them at lunch in the company cafeteria. It may mean a formal thank you letter to a person you interviewed outside the company. It may be recognizing the people who assisted you in your work, even if you can't formally thank them in a publication or presentation. Some typical ways to give credit are these:

- If you're writing a journal article based on your research, thank or at least acknowledge the reviewers and colleagues who criticized earlier drafts of the article.
- If you're writing a book, acknowledge the persons who provided you with special information and served as expert sources of information.
- If you're using someone else's information, get permission and make sure that that permission is published in the appropriate form.
- If you're designing research, make sure that you've conducted a thorough search of the literature for relevant information about similar research and note that research in a literature review.
- If you're citing someone else's information, follow the appropriate style and format for completely and accurately citing your sources.
- If you're the first author or lead researcher on a project, give credit formally and informally to all the other authors or researchers on the project.

- After you've published the results of your research, follow the guidelines that the publisher provided you regarding copyrights. If you want multiple copies of your published work or want to reprint that information elsewhere, you probably will need to get the publisher's approval.

Because technical communicators are also often editors and reviewers of others' work, they may find themselves in the position of seeing researchers' work before it is formally published. For example, an editor of a scientific journal regularly receives original manuscripts from researchers who are eager to publish the results of their research. The editor may assign several SMEs to review the manuscript to see if it provides an adequate description of the research methods and results and if the subject matter is suitable for the specific journal. The editor also reads the manuscript to evaluate the quality of the research, the appropriateness of the subject matter for the publication, and the style and documentation of the research.

Throughout the review process, the researchers who submitted the manuscript assume that their work will be held in confidence, that the reviewers, editors, and others who have access to the previously unpublished research will not publish the research as their own or provide the information to others without giving credit to the researchers. The editors and reviewers who work with unpublished research must also have high ethical standards and established procedures for confidentially reviewing manuscripts and shepherding them throughout the publication process.

Therefore, technical communicators who work primarily or occasionally as editors or reviewers of research publications should follow these guidelines:

- If you're an editor who distributes manuscripts for review, mask the identity of the researchers/authors before the manuscript is first reviewed.
- Require a written statement describing the approval granted for the research and the ethical procedures that were followed while the researchers conducted the experiment or gathered information.
- Require a copyright agreement form that specifies how researchers/authors can use their research after it has been published.
- Keep records of manuscripts submitted for publication, manuscripts going through the publication process, and the copyright and ethical procedures statements signed by researchers/authors.

- As much as possible, verify the accuracy of the information submitted for publication.
- Read all manuscripts in confidence, and keep the information confidential until it is published.

Good research requires teamwork, whether your team involves people you know well and work with daily or people about whom you've only read. It's more than a courtesy, although it is courteous behavior. Ethical behavior in research is certainly part of the research process, but following ethical standards is also important throughout the publication or presentation process, too.

Summary

Ethics is an ongoing part of your professional life. Although ethics is currently receiving a great deal of attention in the workplace and the media, it never goes out of style. You develop and practice the behavior you believe is professional and appropriate for the kind of research you conduct. Acting ethically is a positive, proactive approach to your research. It isn't what you do simply because you don't want to "get caught" or have someone question what you're doing. In this section you've read about several situations that could test your sense of ethics and some strategies for developing your personal code of ethics. The rest is up to you.

Research Activities

1. Critically examine your own research methods and see how you can make yourself more accountable for your findings. How can you improve your documentation of your methods? How can you ensure that your research environment is safe for participants? How can you ensure that your research environment is safe for equipment and materials?
2. If you work with a team on your research project, develop a series of written procedures that you used to ensure ethical research procedures. (If you're working for a business that does not have documentation, like that required for ISO 9000 certification, you may want to develop documentation in light of ISO guidelines.)

3. Find out which companies hiring technical communicators have a formal code of ethics, and read those codes.
4. Talk with members of your institution's ethical standards board or human subjects review board to learn more about the guidelines you're required to follow. Develop a file of the documents you need to submit when you want to gain approval for your research.

For Further Reading

American Psychological Association. 1992. "Ethical Principles of Psychologists and Code of Conduct." *American Psychologist* (December) 47: 1597-611.

American Psychological Association. 1994. *Publication Manual of the American Psychological Association.* 4th ed. pp. 292-97. Washington, DC: American Psychological Association.

Denzin, Norman K., and Yvonna S. Lincoln. (Eds.). 1994. *Handbook of Qualitative Research.* Thousand Oaks, CA: Sage.

Erwin, Edward, Sidney Gendin, and Lowell Kleiman. 1994. *Ethical Issues in Scientific Research: An Anthology.* New York: Garland.

Frederick, William C., and Lee Preston. (Eds.). 1990. *Business Ethics: Research Issues and Empirical Studies*. Greenwich, CT: JAI Press.

Levine, Robert J. 1988. *Ethics and Regulation of Clinical Research.* 2nd ed. New Haven, CT: Yale UP.

Schiltz, Michael E. (Ed.). 1992. *Ethics and Standards in Institutional Research.* San Francisco: Jossey-Bass.

Shrader-Frechette, Kristin. 1994. *Research Ethics.* Lanham, MD: Rowman & Littlefield.

Sieber, Joan E. 1992. *Planning Ethically Responsible Research: A Guide for Students and Internal Review Boards.* Newbury Park, CA: Sage.

Stacks, Don W. 1992. *Essentials of Communication Research.* New York: HarperCollins.

Winkler, Earl R., and Jerrold R. Coombs. (Eds.). 1993. *Applied Ethics: A Reader.* Cambridge, MA: Blackwell.

PART TWO

Ways to Conduct Research

Preparing and planning for research are important parts of the entire process, but the time finally comes when you have to conduct the research you've so carefully planned. Depending upon the type of research you're conducting, whether to write a specific document or to expand the boundaries of knowledge about a subject, you may use one research method or work with several over the course of a project. Part II provides you with strategies for conducting research using several different methods, including both qualitative and quantitative methodologies.

The information among chapters sometimes overlaps, because some methods share similar approaches. For example, one type of questionnaire may be an empirical research tool. When questionnaires are used in this way, your preparation of the questionnaire and some concerns you have about its development and use are similar to concerns you have when you develop a hypothesis and plan an experiment. Another example occurs with the way technical communicators use e-mail in their research. In one of several ways to use e-mail as a research tool, you might write a letter of inquiry via e-mail, instead of as a paper document. Because e-mail has many research uses, you can read about e-mail inquiries in both Chapter 6, Computerized Secondary Sources of Information, and Chapter 9, Letters of Inquiry. Within each chapter, you learn about the source of information and ways to use it, so that you can develop an effective research strategy for any technical communication project.

Part II consists of these chapters:

- Chapter 5, Computerized Information Retrieval from Libraries
- Chapter 6, Computerized Secondary Sources of Information

- Chapter 7, Personal Experience and Knowledge
- Chapter 8, Informational Interviews
- Chapter 9, Letters of Inquiry
- Chapter 10, Questionnaires and Surveys
- Chapter 11, Empirical Experimentation

The first two chapters in this section show you how to locate secondary sources of information by using computerized information-retrieval methods. Chapter 5 deals with indexes, card catalogs, and other typical places where you might find bibliographic information leading you to secondary sources of information. Although the library may be a traditional building on a university campus or in the community, it might be a series of electronic databases to which you gain access through a modem at home, in your car, or at the office, for example. Chapter 6 describes other secondary sources that can provide complete documents, not just bibliographic citations. CD-ROM and network resources, such as mailing lists, discussion groups, multiple-user dimensions, and the World Wide Web, for example, can help you find online journal articles, conference proceedings, graphics, abstracts, e-mail messages, and other secondary sources.

When you need to use primary sources of information, you then use one method or several methods described in the other chapters in this section. Some methods are used more often in one discipline or within one type of technical communication job, but they all reflect research methods technical communicators work with throughout our profession.

CHAPTER FIVE

Computerized Information Retrieval from Libraries

The concept of library research has changed dramatically in the past few years. For generations of students, part of their education came from spending evenings and weekends in the university, college, or community library. There they dutifully searched through volumes of indexes, jotted down the bibliographic listings of information they needed, collected the books and articles, perhaps photocopied the most useful information or checked out materials, and then took notes from the information they read. The research process was a useful part of a person's education, but the time spent in the library was more of an initiation into the realm of scholars, who spent their share of time inside the library.

Research to locate secondary, or sometimes tertiary, sources is still considered a valuable process, one that all students (and all technical communicators) complete when they need background information for a specific project or are learning more about a topic. However, the initiation process into the library research club no longer takes place exclusively in ivy-covered brick buildings. Library research is more likely to take place in a residence hall room, the classroom, at home, in the office, in transit, or anywhere else that researchers can connect to a networked database system.

Some universities are even doing away with the physical buildings of a library, preferring instead to provide students, faculty, administrators, and members of the community with electronic access. As the number of resources increases, so does the need to purchase more resources, store them, and maintain them in good condition. Gaining access electronically to resources located on a network or physically stored in paper form at another location may be the most effective way to try to keep up with the amount of information produced in this Information Age—and to keep down costs at many institutions. The library of the future may be an information infrastructure, rather than a building.

Historically, another common problem with conducting library research may have been limitations to the available sources. The library may not have owned all the sources someone needed, and interlibrary loan took days or weeks. Frequently used sources may have been held in reserve and made available to individuals for short periods of time or only in one place. Other necessary sources may have been checked out of the library for months. Worse yet, some sources may have been damaged or destroyed. Finding important information may have been limited by the sources available in a few libraries.

Today's more common problem is managing the number of sources available electronically. Some estimates suggest that information available through the Internet, for example, is increasing 400 percent each year, as new users gain access to the network and more people are making information available to others through the network. Add to this mix the growing number of commercial networks and services, and you face the probability of being overwhelmed by the amount of information you can access. Without a method of locating what you need and sorting through all the possible sources available to you, you may face new kinds of difficulties.

When you use today's "library" sources, you might work with card catalogs, interlibrary loan, and indexes—but instead of working with them in paper, you're more likely to use them electronically. In this chapter, you'll read about ways you can retrieve computerized information from traditional library devices and services—indexes, catalogs, and interlibrary loan.

To navigate the Information Superhighway successfully, you should be familiar with a variety of tools that will help you gather secondary and tertiary sources of information. These tools include COM and online card catalogs, indexes, and directories on disk—both within companies and institutions and available through networked services. These are just some of the basics. The technology changes daily and affects the ways we work with CD-ROM, networks, and video-based media (through videotape and interactive disks, for example).

Computerized Databases

A *database* is simply a collection of pieces of information that are accessible in many usable ways. Your Rolodex file, phone list, or e-mail address list, for example, are databases that might be in electronic or paper forms. You can find information by name, address, phone number, or some other scheme you've developed.

Technically, just about every library tool you work with is a database or works with a database. A card catalog, whether in paper or electronic form, consists of pieces of information, such as an author's name, the title of a document, the publisher's name, the city of publication, the most recent publication date, and identifying Library of Congress or other cataloging numbers. The drawer or the disk file that contains that information and allows you to access it in several ways, such as by author's name, Library of Congress number, or document title, is a database.

When you work with online databases, you're most likely to work with indexed information that is made available to you through a network. You read abstracts and bibliographic citations that show you where the information you need is located. These abstracts and citations help you determine how much information is available, and of the available information, how much may be useful to you. You download bibliographic information to your e-mail account, printer, or disk. Or you make notes the old-fashioned way—by hand.

Through *local area networks (LANs), wide area networks (WANs),* and the *World Wide Web (WWW),* you locate and retrieve secondary sources to your screen, disk, or printer. If you work with the WWW, for example, you have access to information in multiple media, not just print. You interact with a variety of hypertext or hypermedia formats to learn more about a subject. You upload or download news and comments from bulletin boards and networked files. You subscribe to or just have reader access to electronic journals. As the number and variety of secondary sources continue to increase, you find more potentially valuable information at the click of a mouse or press of a key.

When you work with online databases, you have access to different categories of information that have been compiled to help specific groups of users. For example, you may want to find bibliographic information (a *reference database*) or the original source (a *source database*). However, in most of your searches, whether you use a public domain database, a database provided by a reference service, or a gateway system through which you download information into your computer, you probably will look first at bibliographic information. You might read abstracts and/or citations so that you can easily find the source on your own or can figure out which sources you really need to download or print.

Databases can be stored on tape or disk, most commonly on CD-ROM, but they can also be accessed through a network. Commercial vendors offer a number of databases through one service. If you are a regular subscriber with a vendor, you can do your own searches. If you only occasionally need the vendor's service, you might pay a fee for someone else to search the databases. Because more databases are available through networks that you can access via modem or in a standard location, such as a traditional library, you may not want to have someone else do a search for you.

Other databases may be stored on CD-ROM. Each disk contains lists of sources published within a specific time period. If you want to search for older or newer materials than those listed on the current disk, you have to swap disks. Periodically, the library receives newer disks and makes them available. If you need older disks, you may have to check with a librarian about getting access to them.

Whether the databases you use are available through tape, disk, or network, you need to know how to use them effectively. Having access to all the world's information is worthless if you can't get the information you need in a manageable form when you need it. That's why it's important to develop a search strategy when you want to look up sources about a topic, an author, or a specific publication.

Search Strategies for Using Databases

A database can be an index, a series of documents, a catalog, a directory, or another arrangement of pieces of information that can be retrieved in a variety of ways. You might work with databases through a *gopher*, for example. A gopher is a system to help you work with databases by providing you a menu of options and then carrying out your research request. You might work with hypertext links so that you can find more information about a topic within the database. But however the database is arranged for information retrieval, you'll need to learn more about the kinds of information that are available and the ways you might work with each database to find general or specific information about your subject.

When you need to search for information about a specific topic, you should look for the following types of information:

- how much information is available about a topic
- how much information useful to you for this project is available
- where the information is located
- what type of information is available
- how the information can be accessed

As you seek information about a general subject, you might find hundreds or thousands of database entries that relate, in some way, to that subject. Then your job becomes one of narrowing the search to find just the information that's going to be useful in your current project. By reading bibliographic records stored in the database, you usually learn where the document or material is stored—in which library, and where within that library, for example. You skim several bibliographic listings to learn which types of information are available; you might be looking specifically for conference papers, books, periodical articles, manuals, or reports, for example. The bibliographic information in the database helps you see how many types of information are available, as well as where you might order documents if they're not currently available.

Especially because you're a technical communicator, you should be familiar with several types of online documentation. Online databases most often have good documentation, as prompts or menus on a main screen and through special help documentation that you can call up on the screen when you need it. By reading this documentation, you quickly learn how to access the documents and materials you need, learn more about them, and modify your search strategy to find new information.

If you're researching a "hot" topic or a common subject, you might find hundreds of sources within a single database. If your topic is more obscure, you may have to search many databases to locate all the information you need. In either situation, or more commonly somewhere in the middle, you need to have an effective search strategy in mind to make your search time as profitable as possible.

Your general strategy involves these steps:

1. Develop a list of keywords.
2. Determine which databases are more likely to reference or include the type of resources (for example, conference papers, books) and describe the subject and focus of your project.
3. Refine the search as you learn how many sources are available.
4. Explore other databases to ensure that you've researched the subject thoroughly and possibly to provide you with a new perspective or leads to related subjects.

Develop a List of Keywords

Before you begin your search, you should have a list, either written or mental, of the information you need to locate. If you want to learn about a subject, develop a list of keywords that describe or identify that subject. For example, if you want to learn more about technical communication, your keywords might include *technical communication, technical writing, scientific writing, business writing, scientific communication,* and *technical and scientific communication.*

These keywords would still be very broad, so you might need to determine exactly what you want to know or find about technical communication. You develop another list of keywords that might be used in combination with the first list. You create similar lists for information about authors or titles.

Work with the Appropriate Databases

You should also know where you might logically find information about these keywords. For example, if your emphasis is technical communication, you can probably find information related to technical communication in almost any educational, social science, computer science, communication, humanities, medical, or technological database. By concentrating on databases that cover science, technology, business, and communication, you should find thousands of references to technical communication. At other times, you may find only a few databases that provide you with the information you need. If you need information about the National Science Foundation's (NSF's) grants to individuals, you may only want to search for information in the NSF database.

The emphasis of your research also can help you determine which databases are most appropriate for beginning your search. If your emphasis in a study of technical communication will be on education, you probably first search databases that cover educational resources. ERIC, the Education Index, and other humanities indexes might be good first choices. If, on the other hand, you're interested in technical communication as it relates to the social sciences, you find more appropriate information in indexes like the Science and Humanities Index or the Social Sciences Index.

Refine the Search Strategy as You Work

Armed with a list of keywords and a good idea of which databases to search first, you're ready to begin. However, your search strategy may need further fine tuning as you conduct the search. For example, if you entered the keywords *technical communication* in the ERIC database, you might find thousands of listings (see Figure 5.1).

In fact, this search produced 1,817 items that were cross-referenced by the keyword *technical communication.* That number of entries would be unmanageable, unless you have lots of time to spend reading about sources

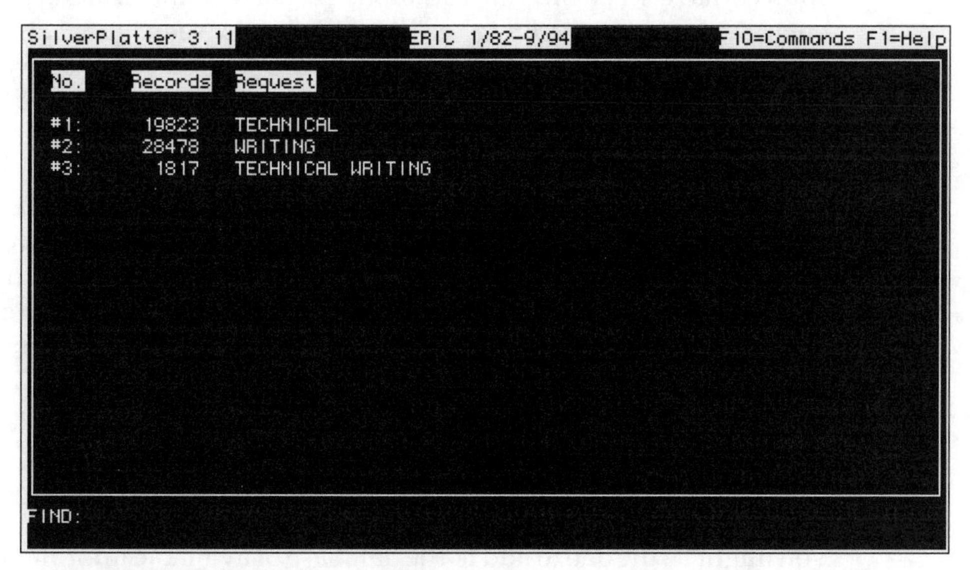

Figure 5.1 ERIC subject search screen.

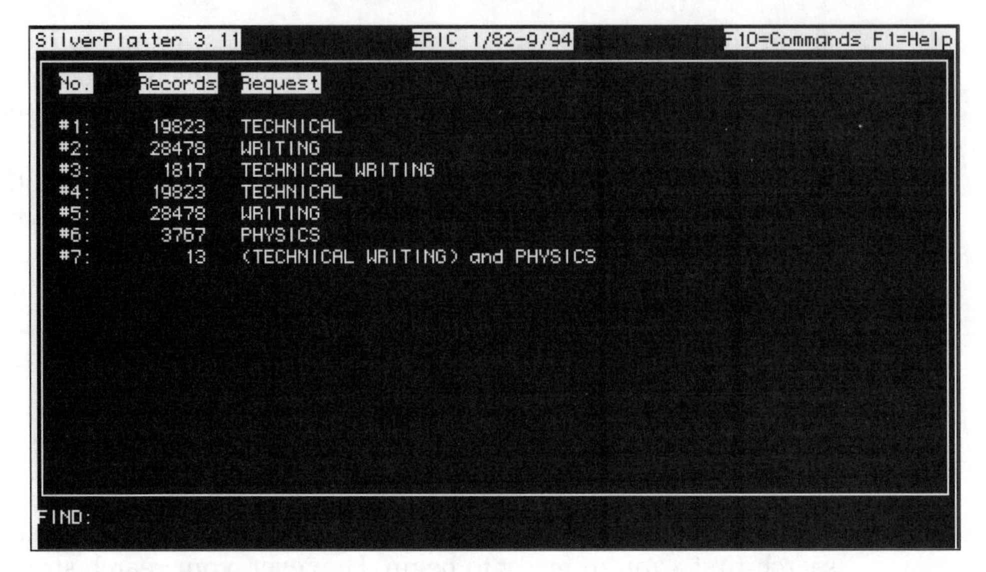

Figure 5.2 ERIC subjects "hits" screen.

that have little to do with your real research interest. Although some entries might be very useful, it would take too long for you to locate them.

If your real research interests involve technical communication as it relates to the theory and practice of physics, you need to see which sources have keywords in common. To narrow the search, you might enter the keywords *technical writing and physics* (see Figure 5.2).

The resulting bibliographic listings are limited to those works that contain information about both technical communication and physics. The number of entries this time is a manageable 13, and you can see all of them.

If, however, the search had produced a number of listings greater than 50, you would need to limit the search further. Perhaps you want only information published since 1993. Depending upon the structure for the search commands used in the current database, you would enter the appropriate command to indicate 1993 as a cutoff date. The resulting number of entries should be more manageable.

The searches you complete can help you in many different ways. You might find that all the sources listed through a search will be useful to your project. You might discover that there are very few sources, but they are all valuable. Or you might learn that because little currently exists, you might be the one to add to the database of available information about this topic; your current project might fill a void in the literature.

Explore Other Databases

To save time, always have an idea of what you need to research and where you might logically find it. The first three steps in your search strategy help you conduct searches efficiently.

Of course, serendipity can play an important part in your ongoing research, and as you have time to explore, you may want to surf the net to explore several databases. You might try working with databases that are new to you, just to see what kinds of information they contain. Because new information and new databases are being made electronically accessible all the time, you might get in the habit of exploring whatever is new to the network. This exploration can pay off rather quickly. You might discover a new perspective on your subject, based on the sources the database includes or the subjects it covers. You might find resources that you didn't know existed, and they will become an important part of any new searches you make. Exploring can be a valuable task, even if you can't map out the territory you want to explore before you begin the exploration. However, if you're under a deadline and need to go directly to information about potential sources for your project, the direct, planned approach is better.

Indexes

An index is a type of database, one in which the fields, a specific type of information (such as author's name, title of a publication, date of publication), and records (i.e., all the information about the same topic, such as a book or a conference paper) are arranged in alphabetical listings. You might search alphabetically through an author index, for example, by the author's last name and, secondarily, by the author's first name. You might search through an alphabetical listing of titles to locate the document you want. When you work with online indexes, you sort automatically these listings in several ways, such as by author's name, date of publication, title of publication, keywords describing the subject, company's name, or publisher's name.

Although some databases networked in a system may contain sources of general information, like a large *Reader's Guide to Periodical Literature,* other databases are highly specialized. Before you use a computerized information retrieval system, you need to understand the type of sources listed in a particular database and the way they're arranged. A library index of sources usually contains information about individual documents, such as reviews, articles, books, conference presentations, papers, dissertations, theses, monographs, and government documents. You can search the database at least by subject, author, or title. Often you can enter keywords as simply a word search, and in some databases you can search by company or periodical. The list of options by which you can search the database usually appears as a menu screen that describes each search method and the way you must structure the search command. By using combinations of keywords, you can learn how many sources are available, where they're located, what they contain, and who created them. You can look for a specific source you already know exists, or you can explore how much information is available, when you may have no idea how many or what kind of sources are available.

Searches by Author, Title, and Subject

In the OhioLINK central catalog, which is available to Ohio universities that belong to a networked consortium of research institutions, researchers can locate information in a variety of ways: by title, author, words, subject, medical subject, call number, or another numeric index. These options are listed in one menu, as shown in Figure 5.3.

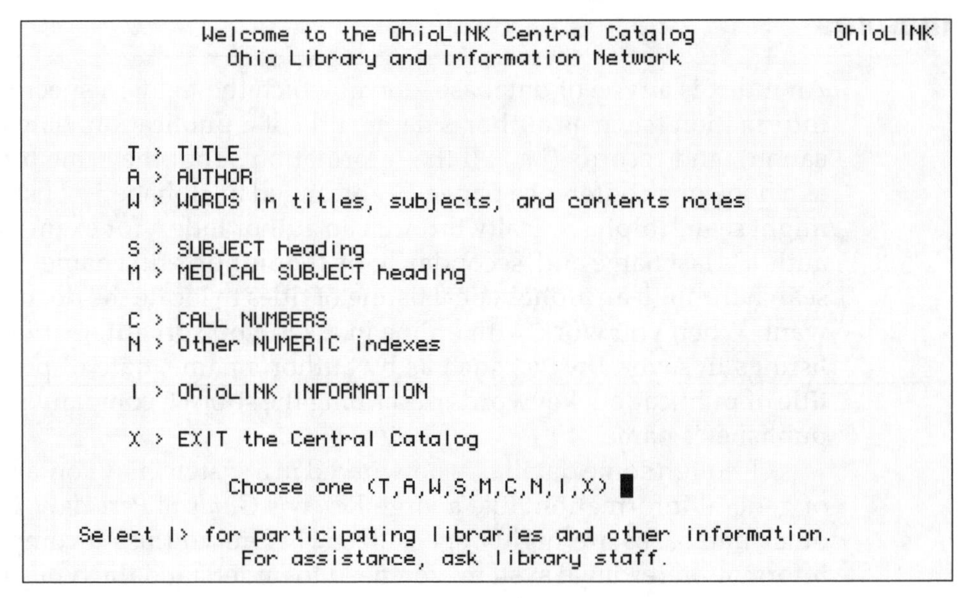

```
          Welcome to the OhioLINK Central Catalog            OhioLINK
            Ohio Library and Information Network

     T > TITLE
     A > AUTHOR
     W > WORDS in titles, subjects, and contents notes

     S > SUBJECT heading
     M > MEDICAL SUBJECT heading

     C > CALL NUMBERS
     N > Other NUMERIC indexes

     I > OhioLINK INFORMATION

     X > EXIT the Central Catalog

             Choose one (T,A,W,S,M,C,N,I,X)

  Select I> for participating libraries and other information.
              For assistance, ask library staff.
```

Figure 5.3 OhioLINK menu.

```
            WORDS  technical communication                         OhioLINK


Type WORDS that may appear in TITLES, SUBJECTS, or CONTENT NOTES:

     for example ---> ADVENTURES OF HUCKLEBERRY FINN

         or just ---> huckleberry finn

     for example ---> teenager or adolescent

     for example ---> letter birmingham jail


                      [Press the ESC key to begin a different search.]

   .... then press the RETURN key
```

Figure 5.4 OhioLINK subject search screen.

If you used the OhioLINK system to search for information about technical communication and ISO 9000, for example, you might first want to conduct a search by (key)words to learn how much information was available. A word search by keywords looks like Figure 5.4.

Searching by words often helps you locate many sources. In this example, however, the combination of keywords produced several entries. The system provides you with information about the number of entries including each keyword in your search and, finally, provides you with the option of seeing the results of this search, as shown in Figure 5.5.

If you knew that the title of a document you needed included the words *technical writing,* but you didn't know if that was the complete title of the work or only part of the title, you might try a search by title (see Figure 5.6).

The resulting title search in OhioLINK produced 79 entries. Twenty documents have the title *Technical Writing.* Other documents have the words *technical writing* within a longer title. You could choose first to look at the 20 entries for documents entitled *Technical Writing* and, if they weren't the document you needed, then proceed to check the other documents (see Figure 5.7).

If you know the name of the author who wrote the document for which you're searching, or if you know that an author has frequently

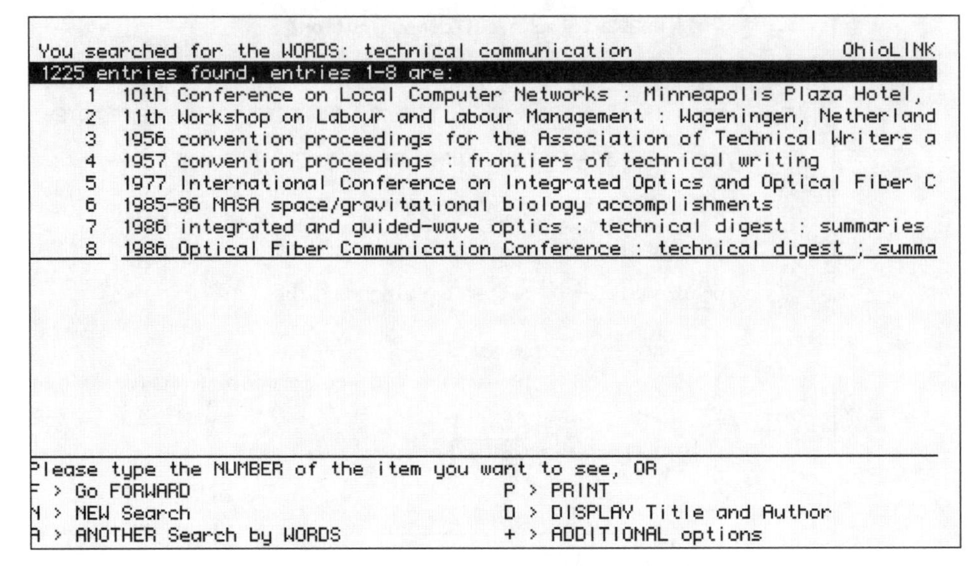

Figure 5.5 OhioLINK subject "hits" screen.

published information about a particular subject, you might search by author (see Figure 5.8).

The example search shown in Figure 5.9 located three entries for what you might assume to be the same author.

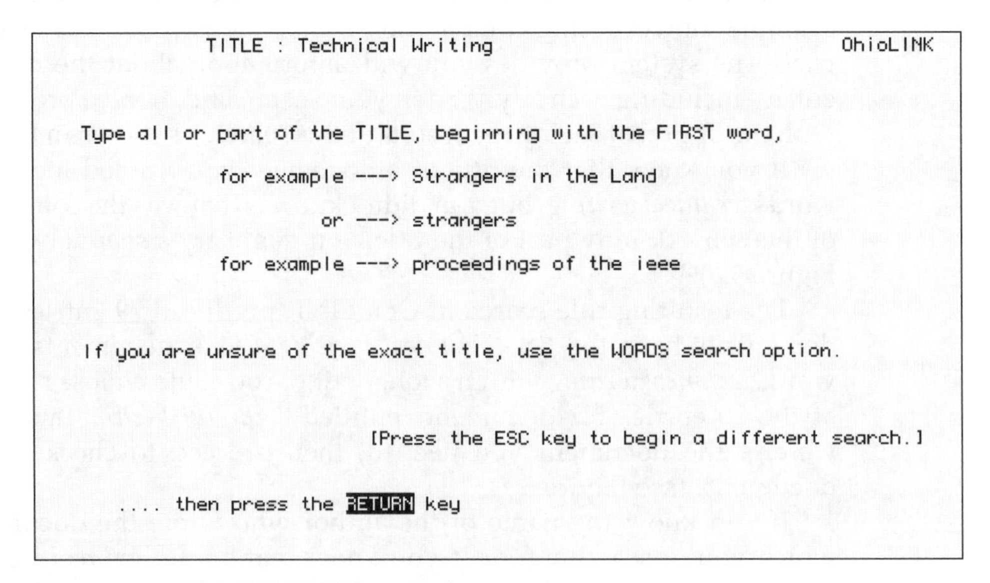

Figure 5.6 OhioLINK title search screen.

```
You searched for the TITLE: technical writing                          OhioLINK
51 TITLES found, with 79 entries; TITLES 1-8 are:

     1  Technical Writing ....................................   20 entries
     2  Technical Writing A Guide With Models ................    1 entry
     3  Technical Writing A Practical Approach ...............    3 entries
     4  Technical Writing A Reader Centered Approach .........    2 entries
     5  Technical Writing And Business Communication .........    1 entry
     6  Technical Writing And Communication ..................    1 entry
     7  Technical Writing And Communications .................    1 entry
     8  Technical Writing And Communications Careers .........    1 entry

Please type the NUMBER of the item you want to see, OR
F > Go FORWARD                              A > ANOTHER Search by TITLE
W > Same search as WORD search              P > PRINT
N > NEW Search                              + > ADDITIONAL options
Choose one (1-8,F,W,N,A,P,D,T,L,J,+)
```

Figure 5.7 OhioLINK title "hits" screen.

Sometimes, if the author's name is common, you might find that information about several different authors results from a single search by author's name. Then check the resulting list of entries very carefully to make sure that you're locating information about the correct person.

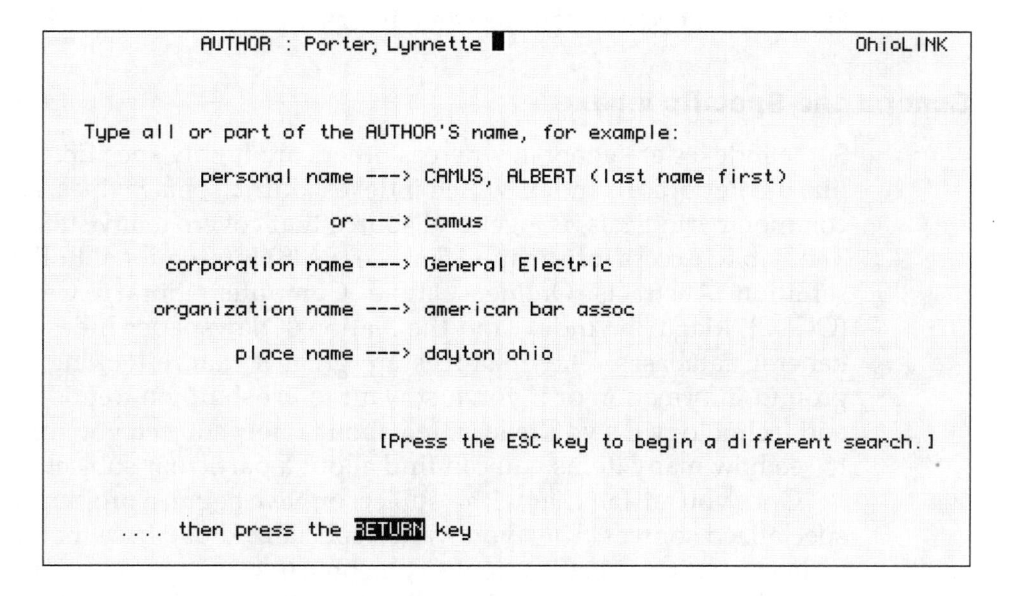

Figure 5.8 OhioLINK author search screen.

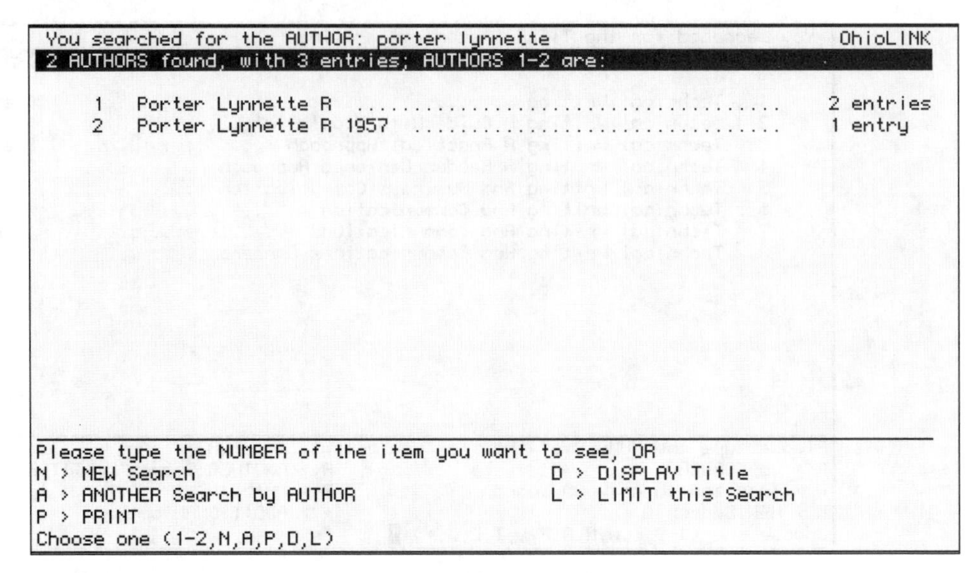

Figure 5.9 OhioLINK author "hits" screen.

Other indexes allow you to search for information through other descriptors, such as publication and company. All the options for searching are usually listed on a menu on one of the first screens you see, so you can quickly learn about your search options and determine the best approach to take for your research needs.

General and Specific Indexes

Some indexes are general, whereas others are highly specific. Databases like the Periodicals Index, which indexes journals, magazines, and other common periodicals, is a general source that covers many subject areas. The Public Access Information Service (PAIS) International Bulletin, Dissertation Abstracts Online, Online Computer Library Center, Inc. (OCLC), Magazine Index, and the National Newspaper Index are other general databases. These sources are great if you're looking for background information or if you just want to brush up on trends in science and technology. As you're learning about a new subject, you might want to see how many items you can find about a particular subject.

Once you are familiar with a subject or have begun a project, you want specialized sources. Then you want a specialized database that lists only sources about a specific scientific, technical, or communication subject. Specialized databases, like the online version of the Science and Humanities Index or the Business Index, provide lists of sources about particular topics in the sciences and the humanities or business. By understanding which index provides you with bibliographic information about the kind of sources you need, you save valuable time in retrieving entries you'll use in your research.

Some indexes frequently found in paper, on CD-ROM disks, or through networks, such as the Internet, are

- ABI/Inform (a business database)
- Applied Science and Technology (an engineering, technology, and science database)
- Art Index
- Biography Index
- Biological and Agricultural Index
- BIP Plus (a database of U.S. and Canadian books in print)
- Book Review Digest
- Cumulative Book Digest
- Dissertation Abstracts
- Education Index
- ERIC (an education resources database)
- Essay and General Literature
- GPO Monthly Catalog (a government documents database)
- Index to Legal Periodicals (a law journal database)
- Library Literature (a library sciences database)
- Medline (a medical database)
- NewsBank (a database listing regional newspapers)
- Newspaper Abstracts Ondisc
- Periodical Abstracts Ondisc
- Philosopher's Index
- PsycLit (a psychology database)
- Sociofile (a sociology database)

These commonly used indexes are often available in community, college, and university libraries. They're also frequently found through educational or research networks.

Other databases that are being accessed more frequently, and thus are found more readily in local libraries, include these:

- AIDSLINE, which covers information about AIDS from 1980.
- CancerLit, which describes cancer therapy and research.
- HEALTH, which describes international literature about health care, planning, facilities, insurance, and other related topics.
- Nursing and Allied Health (CINAHL), which covers English language journals about nursing and the allied health disciplines.
- Anthropological Literature, which lists information about anthropology and archaeology articles and essays.
- Avery Index to Architectural Periodicals.
- Handbook of Latin American Studies, which covers books, journals, and conference proceedings.
- Hispanic American Periodicals Index.
- History of Science and Technology.
- Health Planning and Administration.

These are just a few of the many specialized databases being offered through many networks or services. More entries, more databases, and more links to databases are being established each year.

If you're networked to other databases or vendor services through Internet or Bitnet, or to hypermedia resources through the WWW, you can get access to other libraries', institutions', and companies' collections and databases. Electronic journals and archives, for example, are becoming more readily accessible.

You might want to use navigational interfaces to explore the various databases available to you. Using a gopher is a good way to burrow into new resources. Many gopher systems are set up for groups of special users, such as the members of a university department, a company, or users with special interests in a subject. The gopher then provides menus and documentation that will be appropriate for these people with their special interests. The gopher does the work of completing the tasks to call up the information you've requested and to display what's available. Selecting a gopher from a short list shown in Figure 5.10 is a typical way to begin navigating a network.

This navigational system lets you learn where resources are stored and how to get to them. If you have access to commercial services, such as America Online or CompuServe, or to other services used frequently

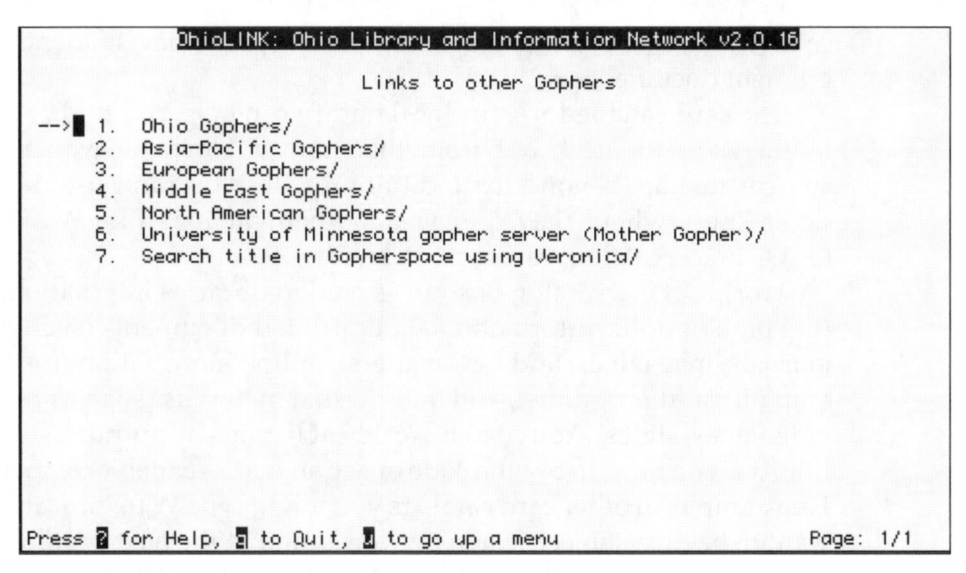

Figure 5.10 Links to other gophers.

by business and academia, such as the Internet, you can work with several gophers. Each one can help you locate information in several different places.

COM and Online Card Catalogs

Most university, college, and larger public libraries use computerized databases to list and track their collections, and even those libraries that are not completely online usually itemize new acquisitions on a database. In any case, whether you're looking for new sources or old faithfuls, most likely you'll work with online "card" catalogs.

Although the information helps you search database listings much faster than you ever could manually, the principles of using an online catalog are similar to those you used when you manually looked up cards in a file drawer. You simply use the library's online card catalog to locate holdings in that library or those available through interlibrary networks. You work with online catalogs for many traditional documents, such as books and periodicals, but use a *computerized output microform (COM)* catalog to find the library's microform collections, often stored on microfiche. Materials stored on microform may include telephone directories, college and university catalogs, company reports, newspapers, and government documents.

The card catalog for your local library can help you find sources you might want to check out from that library. However, you may need sources that are beyond the local library. But you can't use them if you don't know where they are. That's when catalogs like WorldCat, the OCLC union catalog, can come in handy.

WorldCat is a catalog of sources held in libraries internationally. You find bibliographic information about printed documents (such as books, journals, magazines, and newspapers), online information (for example, data files and programs), and audiovisual materials (such as recordings, videotapes, slides). You search WorldCat listings by authors, subject, and title, but you need to use the code or action that's readable by that system. For example, in other card catalogs you might type W to indicate that you want to begin a subject search by words, and at the next prompt type the keyword or combination of keywords to search. When you search for a subject in WorldCat, you type su: and then the keyword(s). Most catalogs, like other databases, provide you with online documentation to help you enter search terms in the correct sequence and with the correct commands.

As with any database, online catalogs list information in several fields, and they usually have familiar field names: author, title, subject. On an additional screen, you may also be asked to supply additional keywords that help you locate subjects when you're not sure of a title or an author (see Figure 5.11).

Next, you read the bibliographic information about the source. Some databases include a table of contents or annotation about the document's content. For example, the bibliographic citation for the source listed through OhioLINK includes the traditional publication information, but it also allows you to look at the table of contents (see Figure 5.12).

Most listings show you where you can find the document and indicate if it is currently available in paper form in the library. You then decide if you want to locate the document, download information, or go on to another source.

Using an online catalog or a COM catalog can be a time saver, but only if you have a good idea of what you're looking for and how you plan to find it. Planning ahead and being familiar with standard keywords save valuable time.

```
You searched for the WORDS: technical communication and ISO 9000     OhioLINK

TITLE        Environmental TQM / John T. Willig, editor.
EDITION      2nd ed.
PUBLISH INFO New York : McGraw-Hill : Executive Enterprises Publications,
               c1994.
DESCRIPT'N   viii, 340 p. : ill. ; 24 cm.
NOTE         Includes bibliographical references.
SUBJECTS     Environmental policy.
             Total quality management.
ALT NAME     Willig, John T.
OTHER TITLES Environmental total quality management.
LC NO        GE170 .W55 1994.
DEWEY NO     363.7 20.
OCLC #       28422931.

------------------------Institutions with this title------------------------
   1> BGU      2> CIN      3> CSU      4> MIA      5> SLO      6> WSU

Key NUMBER to see institution holdings, OR
O > Display OVERVIEW of holdings               R > RETURN to Browsing
K > Display TABLE OF CONTENTS                  N > NEW Search
M > MORE BIBLIOGRAPHIC Record                  + > ADDITIONAL options
Choose one (1-6,O,K,M,R,N,A,S,P,T,E,+)
```

Figure 5.11 Database annotation screen.

```
You searched for the WORDS: technical communication and ISO 9000     OhioLINK

TITLE        Environmental TQM / John T. Willig, editor

Introduction / John T. Willig

Ch. 1 TQM and Strategic Environmental Management / Gene Blake ............  1
Ch. 2 Corporate Environmental Management Survey Shows Shift from
        Compliance to Strategy / Tony Lent, Richard P. Wells .............  7
Ch. 3 Certification of Environmental Management System - ISO 9000 and
        Competitive Advantage / Suzan L. Jackson ......................... 33
Ch. 4 A Point of View: Seven Principles of Quality leaders / Y.K. Shetty .. 43
Ch. 5 The Team Approach to Companywide Change / Robert Janson, Richard L.
        Gunderson ........................................................ 51
Ch. 6 Hitting the Wall: How To Survive Your Quality Program's First Crisis
        / Barry Sheehy ................................................... 63
Ch. 7 Environmental Leadership Plus Total Quality Management Equals
        Continuous Improvement / Abhay K. Bhushan, James C. MacKenzie ..... 71
Ch. 8 Environmental Marketing's New Relationship with Corporate
        Environmental Management / Walter Coddington ..................... 95
F > FORWARD                                  A > ANOTHER Search by WORDS
R > RETURN to previous screen                P > PRINT
N > NEW Search
Choose one (F,R,N,A,P)
```

Figure 5.12 Database table of contents screen.

Interlibrary Access

If your home library lacks the information you need, you may be able to locate documents by author, subject, or title in another library's catalog. Many larger universities' libraries are linked via computer networks as part of an online consortium. If you have access to the consortium's network, you locate the sources you need from any one of the member libraries, even though you're probably separated by miles.

For example, if you want to look up possible sources about research grants and you have access to the Internet, you might use one of several gophers to search for available information. One consortium's system, the OhioLINK gopher, links private and public colleges, universities, and technical schools with the state's library. OhioLINK allows users to search through the computerized catalogs at several university libraries. Other states and consortia have established similar networks, through which you find information about sources located in any of the participating libraries. You find that several libraries have books about research grants, and you can learn how much it will cost if the book is sent to you.

OCLC provides an electronic communication system among libraries. It, too, can help you locate sources kept at other libraries, no matter how far they are from you geographically. This type of interlibrary loan access used to be available only to those librarians who checked for documents for you. However, with the increasing availability of networked libraries—as well as other repositories of information, you often can find the sources yourself.

If you find that the source you need is in another library, either you directly or with the assistance of a librarian in your home library can order the source online. Because you can tell from the bibliographic information if the source is checked out or is "on the shelf" before you make the request, you know where and if the sources you'd like to borrow are available. Sources like books, for example, still have to be sent to you. Other sources, such as articles from periodicals, may need to be photocopied, or they may be downloaded to you electronically.

Summary

Computerized information retrieval constantly updates the traditional ways of gathering information and determining the scope of a research topic. You could spend hours or days browsing through databases. But much of that time wouldn't be productive. To be an effective researcher, you need to be able to sift through the increasing amount of electronic information, locate those pieces of information that are useful, and keep learning about newer sources of electronic information. You can download information to a printer, a disk, or an e-mail account so that you can create a working bibliography of available sources. You can order information if you need the hardcopy document. Computerized information retrieval in libraries provides you many services and ways to gain access to the materials you need for a project.

Computerized information retrieval gives you access to hundreds of sources in different media—sources that you might not have been able to find until the information was networked. But electronic media also give you access to people, and the skills you develop for interviewing, writing, and editing are increasingly important when you work online.

Research Activities

1. Find out which databases are available to you through your company, local libraries, universities, and colleges.
2. Learn how you can get an e-mail account. Determine which networks are available for you through this account.
3. Using library sources, compile a working bibliography of at least 25 sources that will provide you background information for your research project. Download documents that you might not find in hardcopy form in your library. Download the citations you've located onto your own floppy disk or online files.
4. Spend some time "surfing" or browsing through different databases. Experiment with them to learn what type of information is available; how it can be accessed; if it can be downloaded into an e-mail account or onto a disk, as well as to a printer; and how to work with the system.

For Further Reading

The following sources are only a few representative examples of the variety of information available about databases and online information retrieval. More sources about the use of databases, indexes, catalogs, and other online access to information are being published each month.

Curtis, Donnelyn, and Stephan A. Bernhardt. 1991. "Keywords, Titles, Abstracts, and Online Searches: Implications for Technical Writing." *Technical Writing Teacher* (Spring): 142-61.

"Databases: Some Options for End-user Searching in the Health Sciences." 1989. *Reference Quarterly* (Spring): 395-406.

Gardner, Sylvia A. 1992. "Spelling Errors in Online Databases: What the Technical Communicator Should Know." *Technical Communication* (February): 50-3.

Schwarzwalder, Robert. 1994. "Searching at the Fringes: Finding Technical Information in Nontechnical Databases." *Database* (April): 100-04.

Stephens, Irving E. 1991. "Citation Indexes Improve Bibliography in Technical Communication." *Journal of Technical Writing and Communication* 117-25.

CHAPTER SIX

Computerized Secondary Sources of Information

Computerized databases made available on CD-ROM and through networks are good places to find bibliographic information that can direct you to secondary sources. Learning which sources are available and where they're located is an important part of your research process. Sometimes, however, you need immediate access to the source, and you don't want to visit the nearest library or spend days waiting for materials to arrive through interlibrary loan. Fortunately, you can now directly access some secondary sources. Computerized secondary sources found in source databases help you quickly find and compare information online. Some source databases provide complete newsletter, newspaper, or journal articles and conference papers. Reading the information online saves you time in locating the paper copies of secondary sources, perhaps photocopying them or checking them out of the library, and then reading them. Only if you must have a printout will you download the information from the source database and print it.

Electronic mail systems provide you with another way of retrieving secondary sources of information. E-mail provides you with informal, current secondary sources—messages from colleagues, experts, and people with similar special interests. Through e-mail, you collaborate with other technical communicators, ask questions of technical or scientific specialists, and share your ideas about specific projects or general subjects. Computerized information retrieval means establishing a link with people, as well as materials, to help you learn more about a subject.

In this chapter you'll read about some computerized secondary sources of information you can access in their complete form: articles, abstracts, conference papers, e-mail messages, and multimedia materials. A source of information you might use daily is e-mail, as you collaborate with colleagues and request advice and information. You might also discuss your research, learn about topics and events that might have an impact on your work, and simply share ideas, even playfully, with colleagues through bulletin boards, newsgroups, discussion groups, and multiple-user dimensions. When you need more formal, or less personal, research, you might work with information that's already been placed in a networked database. As more information is placed on networks, especially the World Wide Web (WWW), you have more access to complete secondary sources from anyplace where you can establish a network link. You also have more opportunities to work with CD-ROM-stored materials.

E-mail in Research

E-mail seems to be everywhere these days, and if you enjoy rapid electronic communication with colleagues, friends, and strangers who let you in on their ideas, then e-mail is probably a primary means for you to gather information. But if you've logged onto your e-mail system only to find thousands of messages, many of them junk mail, clogging your and your computer's memory, then you may not be so happy with the e-mail revolution.

When it's used properly, e-mail can be an important secondary source of information. It lets you communicate immediately with people who might have answers to your questions or be able to provide information or direction for your research. It may provide you with ideas for future research projects. It may give you a group of people from whom you can elicit more formal responses, such as in surveys or online questionnaires. You can read more about e-mail as a less formal way of sending letters of inquiry in Chapter 9, Letters of Inquiry.

E-mail surveys, usually sent through a mailing list that reaches hundreds of e-mail addresses daily, are becoming more common. If you decide to poll the users of an e-mail system, you might provide a very specific subject line, so that potential respondents know that the message contains a survey about a specific topic. Then, when you provide the survey, be brief. Your e-mail format should be easy to read and use, and you should limit the amount of information. Just like respondents may not complete a questionnaire if they have to read page after page, so e-mail respondents may not have time to read screen after screen of information. However, because e-mail provides an immediate forum in which to respond, researchers frequently find that they get many responses to an e-mail survey. (More information about survey or questionnaire design is provided in Chapter 10, Questionnaires and Surveys.) Although the medium for sending and receiving responses is different than that used with paper questionnaires, the design concepts are very similar.

Strategies for Using E-mail in Research

Using e-mail as a communication and a research tool requires a few guidelines. If you practice good netiquette, you'll find that e-mail will be both interesting and profitable. Here, then, are the three Rs for your electronic education and enjoyment:

- Read the important messages.
- Respond politely, succinctly, and promptly.
- Request information politely, succinctly, and specifically.

Read the subject lines, if they're available, to determine the priority of reading your mail. If you subscribe to lists and bulletin boards, you may find hundreds of messages daily posted to your e-mail address. Get in the habit of scanning your directory or list of messages and prioritize your mail. Then, if you have the time and inclination to read the general messages, philosophical posts, and personal chitchat, you can get to them after the important mail has been taken care of.

Before you forget or get too far from your e-mail, respond to the requests that have been made of you. Be direct and friendly, and provide the information that was specifically requested. Although e-mail may seem remote and impersonal as you and your keyboard respond to a faceless message sender, remember that you'll soon be one of those message senders, and you want your message to be a good reflection of you, your knowledge, and your pleasing personality.

Finally, when you do need to send a request, make sure that you know to whom you want to send the request. Postings to a mailing list may sometimes be appropriate, if you want lots of feedback to your request or question, but most often you'll want to ask a few people for information. Ask for specific information or pose a specific question.

A good way to alert someone that your message is important and contains a request is to write a carefully worded subject line. A subject line serves a number of functions. It helps readers prioritize their messages. (And if you're making a request that you hope will be complied with immediately, you want your message to have a high priority.) A subject line indicates if the message's subject matter will be interesting to them or if the message provides important information they can use that day. When you submit a request, add an informative, concise subject line that describes the subject of your request.

When you write any e-mail message, use correct grammar. Remember that you're a technical communicator—and presumably you can write and edit well. Your e-mail messages shouldn't be exempt from your good writing and editing. Keep your message focused and polite. Unless you enjoy flame wars with other e-mailers, you want to write thoughtfully, not out of pique. Your signature should identify you, your address, and perhaps a logo representing you or your business. Try to avoid cutesy sign-offs and smilies that take up space and add little to your message or reputation. For more information about writing effective e-mail messages of request, you should study Chapter 9, Letters of Inquiry.

Bulletin Boards

Along with e-mail, bulletin boards are increasing in number and popularity. Like traditional bulletin boards, the electronic version features notices about professional meetings, job announcements, and other short news items. The information can be uploaded or downloaded with the permission and instruction of the group or company sponsoring the bulletin board. The information on a bulletin board listing is usually just a starting point, but from that point, you can find other sources of information.

The Society for Technical Communication (STC) provides one bulletin board for technical communicators. If you want to post a notice, you need to meet the criteria for postings set by the Society. But the technical requirements for accessing the bulletin board are minimal, and, if you're an STC member, you can get instructions for using the bulletin board. You might call or send an e-mail message to STC's international office in Arlington, Virginia, for more information about their bulletin board and procedures for posting notices.

Newsgroups, Mailing Lists, and Discussion Groups

If you have a special research interest, hobby, business-related interest, or educational need, you can probably find a newsgroup, mailing list, or discussion group made up of people who share that interest. For example, people who subscribe to the techwr-list mailing list receive e-mail messages sent to anyone on that list or to the whole list in general. On a given day, hundreds of messages may be sent to newsgroup members. Some messages are part of a running dialogue among newsgroup members. Some messages may be requests for information or solicitations for participation in an e-mail survey. Some messages may be humorous asides or serious commentaries. But they all relate in some way to technical writing and should be at least minimally interesting to the technical writers who subscribe to the list. Members of the SGML mailing list often find helpful suggestions or shared concerns about using SGML in the workplace. Subscribers often describe work experiences and pass along information that will be helpful to anyone who works with SGML. These are only two examples of the growing number of mailing lists that can help technical communicators learn more about specific topics or share general information.

Subscribers to or participants in a group can lurk on the line and passively read messages, or they can send their comments, inquiries, and suggestions to other members. If you do participate in a group, you might want to read several messages or observe the group's protocols first to see what people are discussing and doing.

There are hundreds of newsgroups and mailing lists, dealing with all kinds of topics, from business-related matters to entertaining pastimes. To get an idea of some possible lists, turn to Appendix B, an abbreviated list of some currently available newsgroups. Discussion groups may be more focused on a particular topic, at least for a session, or they may be an event, almost like a party to which the participants can bring any discussion topics of interest. Discussion groups may be by invitation only or announced through e-mail or a bulletin board.

If you really like to participate in discussions, you may want to try a multiple-user dimension (MUD), a virtual reality environment in which you can play a role and interact with other characters. Many MUDs are games, with characters like monsters, knights, and gnomes, but a MUD can be as serious or as playful as its participants like. MUDs are fun to explore when you have free time, and as a technical communicator, you probably will be interested in seeing how people communicate in virtual reality. However, for the nuts and bolts practical research you do on the job, probably a MUD is better reserved for free time on the net.

Electronic Journals

In addition to finding out about sources of information that you can have sent to you, download electronically, or print, you can read some secondary sources directly online. Several gophers lead you to electronic journals, bulletins, and notices stored in databases. You can read the original copy, instead of finding an abstract or a citation indicating what kind of information is available and where it can be found.

Many professional societies, in addition to traditional publishers of consumer publications, have entered the electronic publication market. Some publishers provide an electronic as well as a hardcopy volume of each issue. Just a few scientific electronic journals currently accessible via Internet are the *Journal of the International Academy of Hospitality Research, Chicago Journal of Theoretical Computer Science,* and the *Journal of Counseling & Development.* A sample listing of electronic journals or gophers leading to journals and notices is found in Figure 6.1.

Not all back issues may be available electronically, and many publications are scrollable only—they are not linked with hypertext words or icons. However, the sophistication of electronic journals is increasing, and the number of journals going electronic has more than doubled within a year. This trend in electronic publishing is just beginning. As more people have access to electronic journals, and as the need for immediate access to journals increases, the electronic format should gain popularity.

Most online journals provide you with a table of contents and a separate list of articles by title, which you can then retrieve and read or otherwise store for future reference. The online articles may take awhile to retrieve, depending upon the size of the file, and you usually have to scroll through the article. Still, the information is much more readily

```
OhioLINK: Ohio Library and Information Network v2.0.16

                    Electronic Journals

 -->  1.  Academe This Week/
      2.  CICNet Electronic Journal Project (full-text)/
      3.  CONSER, serials catalog (via CARL) <TEL>
      4.  Newspapers, Magazines, and Newsletters/
      5.  Electronic Newstand/

Press ? for Help, q to Quit, u to go up a menu              Page: 1/1
```

Figure 6.1 Ohio Library Electronic Journals screen.

available than if you had to go to a library and retrieve a paper copy of the journal.

The first page of a recent *Chronicle of Higher Education* article, available in its entirety through an Internet database is shown in Figure 6.2. The

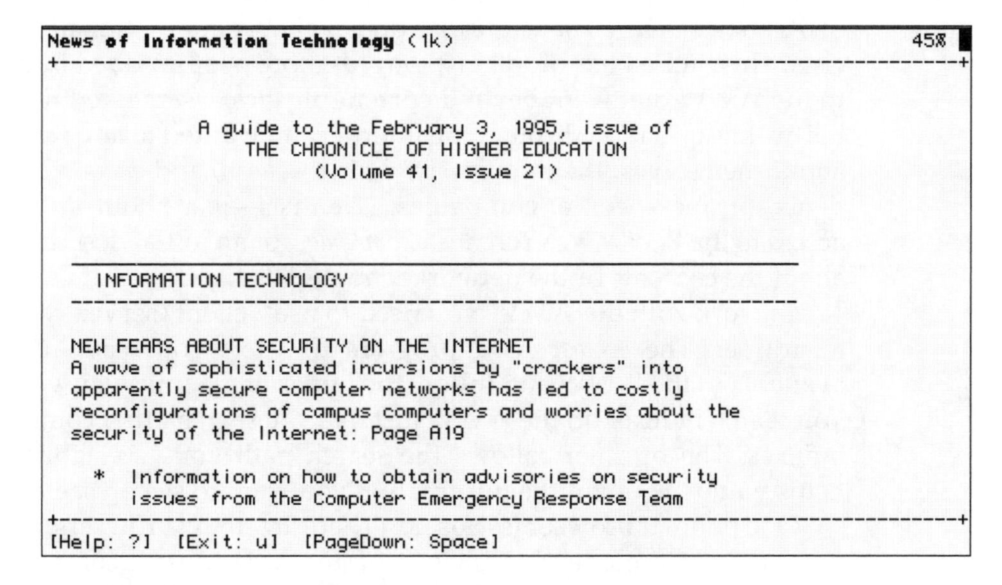

```
News of Information Technology (1k)                              45%
+--------------------------------------------------------------------+

          A guide to the February 3, 1995, issue of
              THE CHRONICLE OF HIGHER EDUCATION
                    (Volume 41, Issue 21)

  ____________________________________________________________
    INFORMATION TECHNOLOGY
  ------------------------------------------------------------

  NEW FEARS ABOUT SECURITY ON THE INTERNET
  A wave of sophisticated incursions by "crackers" into
  apparently secure computer networks has led to costly
  reconfigurations of campus computers and worries about the
  security of the Internet: Page A19

     *  Information on how to obtain advisories on security
        issues from the Computer Emergency Response Team
+--------------------------------------------------------------------+
[Help: ?]  [Exit: u]  [PageDown: Space]
```

Figure 6.2 Online version of *The Chronicle of Higher Education*.

article is scrollable, but documentation on the screen lets you know how many "pages" (usually a screen's worth of information) are in the article and where you are within the document.

Even popular magazines like *People* are making the information they publish available in print and online. Many publishers expect that people who formerly were readers are becoming browsers or interacters with their information. Making information, whether for scholarly research or general interest, available in interactive formats helps network users gain immediate access to constantly updated information.

The World Wide Web

One of the most exciting online services is the World Wide Web (WWW). Sometimes it's referred to as the World-Wide Web, 3W, or just the Web. Regardless what it's called, it's an exciting place to work. The WWW provides you with hypermedia information with several links, so you can move from place to place within the material, according to your interest in the information. The WWW is one way to organize information available through the Internet and make it accessible to more users internationally.

Originally, the WWW was developed at the CERN research center in Switzerland to help make information available internationally to physicists. This limited use didn't last long, as more people added information to the WWW that went beyond current physics. Users liked the accessibility of information through the Internet and wanted to add information about themselves, their company, their research, and so on. Today, students, businesspeople, companies, scientists—just about anyone with access to the WWW—can create a *home page*, or an initial descriptive page about the contents of the networked materials.

To work with the WWW, you need to use a client/server system with a browser. The *browser* is a program that lets you read information available in the WWW and follow the hypermedia links you've selected. You search indexes on the WWW just like you search other computerized indexes. The browser follows the search path you've established and helps you track down the information you want to see on screen.

One popular browser is Mosaic. (There are many browsers, and we're not trying to promote one instead of another.) Mosaic comes from the National Center for Supercomputer Applications (NCSA) at the Univer-

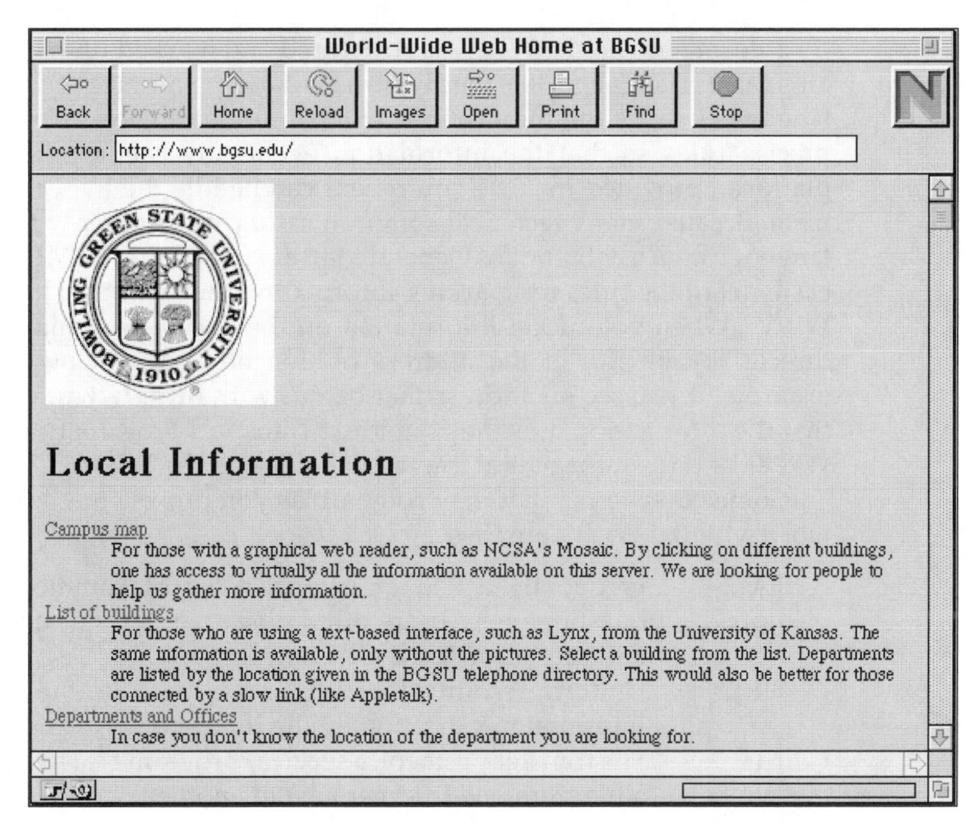

Figure 6.3 Bowling Green State University World Wide Web home page.

sity of Illinois at Urbana. Mosaic formats hypermedia information so that you can easily use it and keep track of the links you've followed.

The WWW is becoming an increasingly important research tool for technical communicators. In case you haven't accessed information on the WWW, Figure 6.3 shows a fairly typical home page and a sample of the information you might find. On a color monitor, you would see that the hypertext links are underlined in one color. Then, when you follow a link, the color changes to indicate that you've used that link.

Not all researchers have access to the WWW, because it may require a faster modem, color monitor, or browser that they don't have. However, more people with e-mail connections to the Internet are gaining access to the WWW, and installing the appropriate browser for your system and upgrading your connections to take advantage of the WWW is a worthwhile investment.

Information on the WWW is structured and marked up with a standard markup language, like the Hypertext Markup Language (HTML). The language's codes and the structure of the information being coded follow precise rules so that the information eventually can be read by any platform, provided that the equipment can handle the hypermedia formats. Because many technical communicators are familiar with markup languages, such as Standard General Markup Language (SGML), they can easily learn the rules for marking information they want to share via the WWW. When you locate information on the WWW, you may access it directly by keying in the address of the information you need. For example, if you see an address that begins with **http://www.**, you know that the information uses the Hypertext Transport Protocol (http) for the WWW, and the information is available through the WWW.

Some other terms that may come up as you browse the Internet and work with the WWW are these:

- FAQ — Frequently asked questions, a list of commonly asked questions new users often pose to a newsgroup, and the answers provided by the newsgroup.
- URL — Uniform resource locator, the technical description of the information's location on the Internet.
- FTP — File transfer protocol, a service/program that lets you copy a file from one Internet host to another.

One introductory FAQ screen for the Veronica gopher is shown in Figure 6.4. You might want to read these files first when you work with gophers and the WWW.

CD-ROM Information

Some periodicals, such as journals and newsletters, may regularly be presented online, but another electronic format for periodicals is CD-ROM. Conference proceedings, for either a single conference or multiple conferences, are occasionally offered on CD-ROM, as well as on paper. Some journals are archived, so that a year's issues appear on one disk. Dissertations and theses may be archived this way, too.

Most periodicals stored on CD-ROM have a document resembling a table of contents, from which you might be directed to several indexes, such as subject or topic, author, and title indexes, just as you use in working with reference databases. You might search an index to learn, in

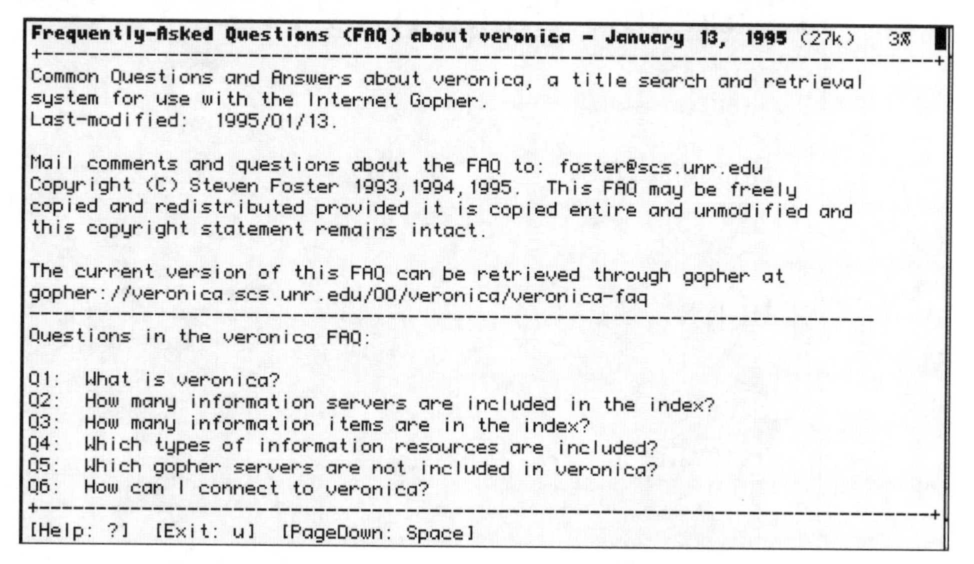

```
Frequently-Asked Questions (FAQ) about veronica - January 13, 1995 (27k)   3%
+-----------------------------------------------------------------------------+
Common Questions and Answers about veronica, a title search and retrieval
system for use with the Internet Gopher.
Last-modified:  1995/01/13.

Mail comments and questions about the FAQ to: foster@scs.unr.edu
Copyright (C) Steven Foster 1993,1994,1995.  This FAQ may be freely
copied and redistributed provided it is copied entire and unmodified and
this copyright statement remains intact.

The current version of this FAQ can be retrieved through gopher at
gopher://veronica.scs.unr.edu/00/veronica/veronica-faq
------------------------------------------------------------------------
Questions in the veronica FAQ:

Q1:  What is veronica?
Q2:  How many information servers are included in the index?
Q3:  How many information items are in the index?
Q4:  Which types of information resources are included?
Q5:  Which gopher servers are not included in veronica?
Q6:  How can I connect to veronica?
+-----------------------------------------------------------------------------+
[Help: ?]  [Exit: u]  [PageDown: Space]
```

Figure 6.4 Veronica FAQ.

general, which articles match your research interests, or you can locate a specific article by its title or author. Many disks provide hypertext links so you can quickly access information without having to scroll through a long document.

One recent example of a CD-ROM disk containing proceedings from a group of related conferences was designed by the International Society for Optical Engineering, known as SPIE. Each of the proceedings is structured so that researchers can use hypertext links to move from one menu layer to another, one part of the proceedings to another, and one type of documentation to another. Although the CD-ROM format is more expensive than traditional paper, the proceedings on disk provide ready access to several volumes and allow quick searches for information throughout the proceedings. Figure 6.5 shows the table of contents for one volume of the SPIE proceedings stored on CD-ROM.

From this point, you could choose where to look for information. If you decided to look through titles in a volume of the proceedings, you might search by title. A search by the keyword *hologram* from the alphabetical list of title keywords brought up seven "hits." From the scrollable list of proceedings papers' titles, you could choose the title(s) that seemed most appropriate for your research, as shown in Figure 6.6.

If you're not sure which articles you want to read in their entirety, you might read several abstracts to determine if the approach and subject

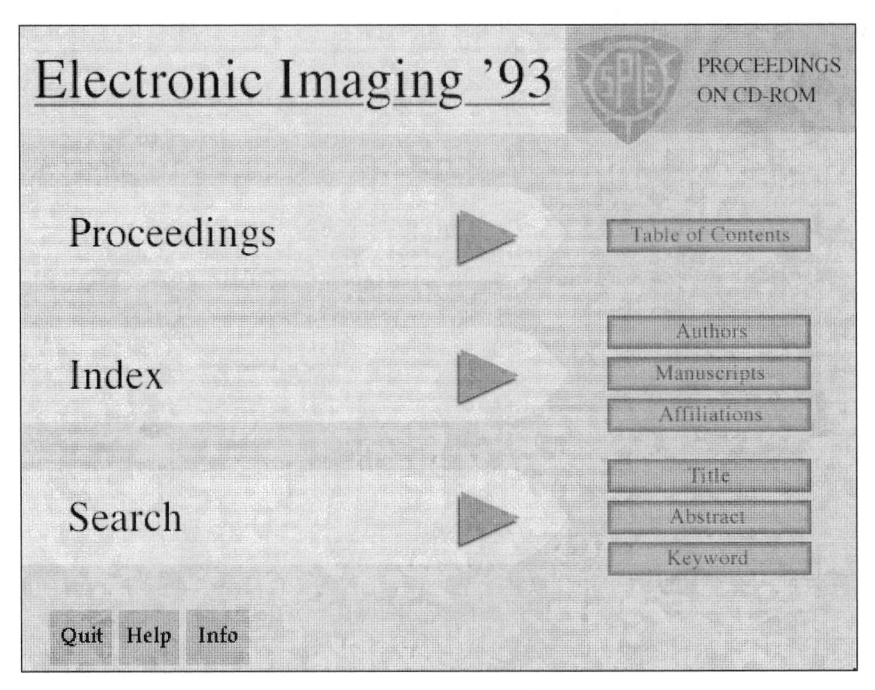

Figure 6.5 SPIE Table of Contents screen.

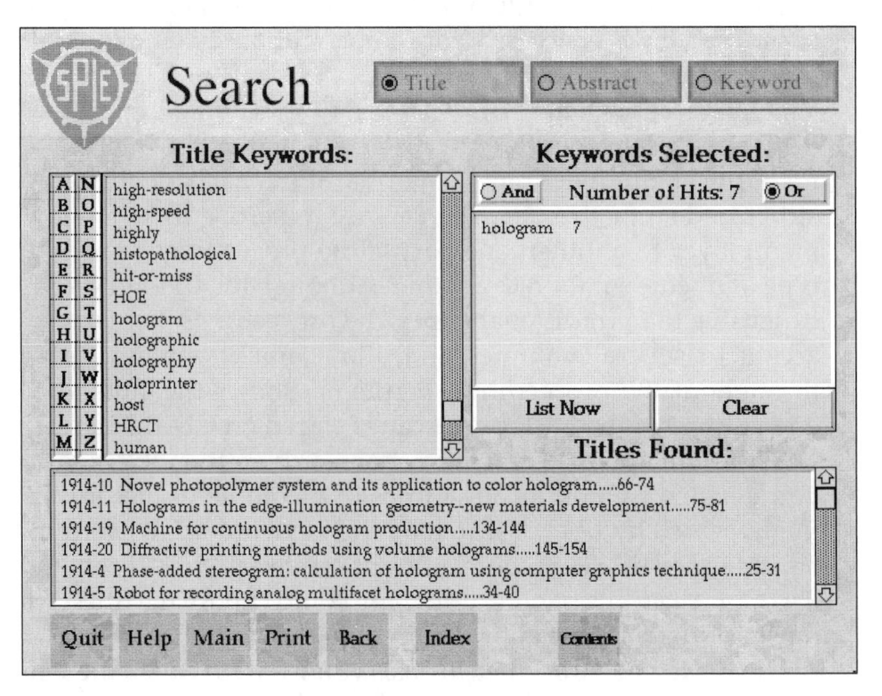

Figure 6.6 SPIE Search by Title Words screen.

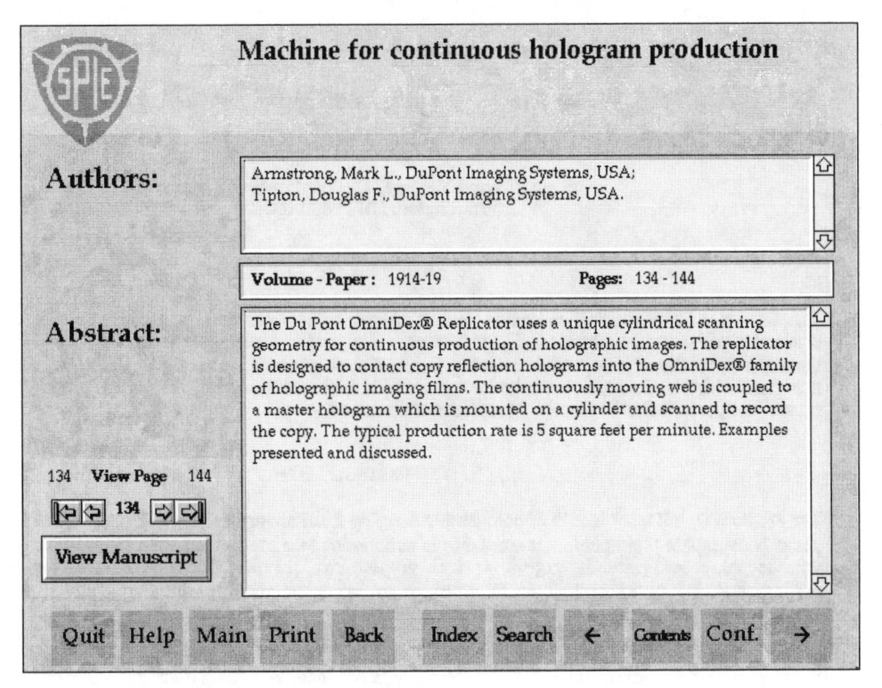

Figure 6.7 SPIE Abstract screen.

matter in an article meet your research needs. If the abstract provides helpful information, you can then read the full description of the research in the paper. An abstract for one proceedings paper summarized the research relating to continuous hologram production (see Figure 6.7).

The text of the conference paper is scrollable. A "floating" palette contains buttons that help you move forward or backward one screen or to the beginning or end of the paper, as well as print the paper or choose from among other text-management options. A first screen of the paper that was summarized in the preceding abstract is shown in Figure 6.8.

Well-designed CD-ROM collections of documents provide you with several hypertext or hypermedia links so you can move quickly among articles, menus, indexes, tables of contents, Help documentation, and other information available on the disk. Clickable icons as well as scrollable screens make the information easier to skim and help you locate information that's most important to you.

Most disks containing periodicals, such as conference proceedings, are purchased and used by regular subscribers or members of a professional association that published the periodicals. They are not currently

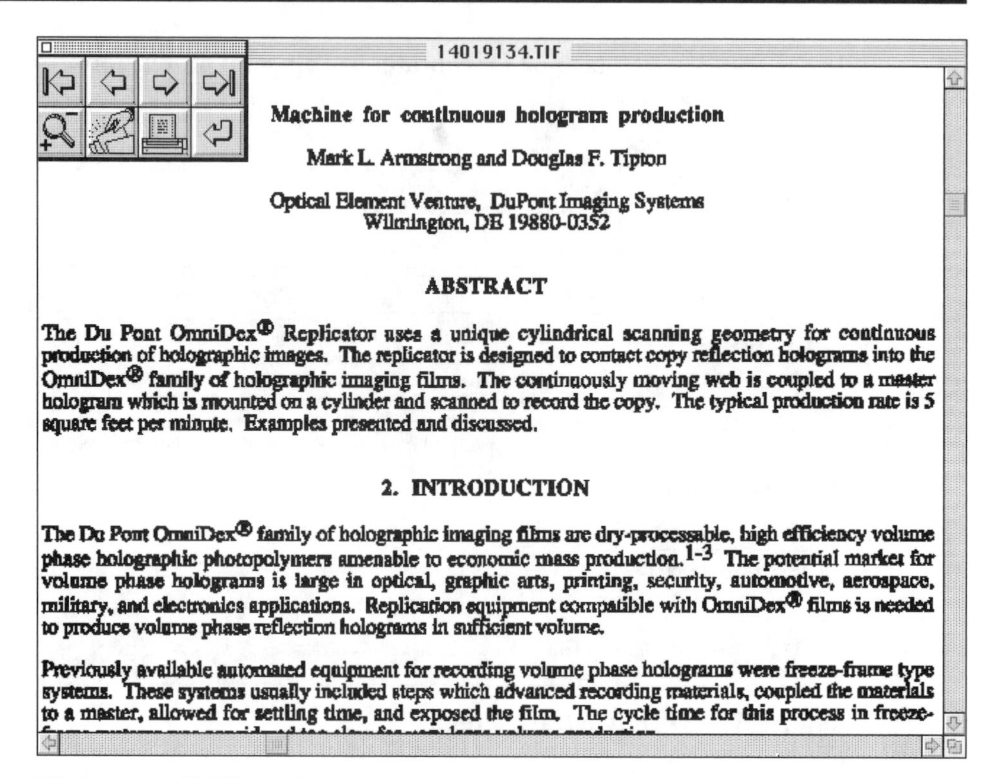

Machine for continuous hologram production

Mark L. Armstrong and Douglas F. Tipton

Optical Element Venture, DuPont Imaging Systems
Wilmington, DE 19880-0352

ABSTRACT

The Du Pont OmniDex® Replicator uses a unique cylindrical scanning geometry for continuous production of holographic images. The replicator is designed to contact copy reflection holograms into the OmniDex® family of holographic imaging films. The continuously moving web is coupled to a master hologram which is mounted on a cylinder and scanned to record the copy. The typical production rate is 5 square feet per minute. Examples presented and discussed.

2. INTRODUCTION

The Du Pont OmniDex® family of holographic imaging films are dry-processable, high efficiency volume phase holographic photopolymers amenable to economic mass production.[1-3] The potential market for volume phase holograms is large in optical, graphic arts, printing, security, automotive, aerospace, military, and electronics applications. Replication equipment compatible with OmniDex® films is needed to produce volume phase reflection holograms in sufficient volume.

Previously available automated equipment for recording volume phase holograms were freeze-frame type systems. These systems usually included steps which advanced recording materials, coupled the materials to a master, allowed for settling time, and exposed the film. The cycle time for this process in freeze-

Figure 6.8 SPIE conference paper.

available in most libraries; however, they can be a good portable source of information.

Strategies for Using CD-ROM or Network Research Tools

Searching for information stored on CD-ROM or on the WWW is not much different from using any other computerized database. You still need to determine a list of search terms and have an idea of where you want to search. By working with the menus provided by the browser or gopher, you select the way you want to search the database and where you want to look.

Working with hypermedia or hypertext links can both be more complex and more fun, because they allow you to move according to your interests, intuition, or inclination among types of information or within a document. You might click on an icon just to see where you go or what kind of information is presented. You might click a highlighted term to learn more specifically about that topic or to find related information. You might look at different types of graphics, in addition to text.

Many systems or browsers allow you to see which hypertext or hypermedia links you've already used. For example, Mosaic highlights in color those links that you've already used. Other systems may provide a tree structure showing the search path you've taken. You might want to retrace your path or try another path altogether. In a well-designed system, you won't get lost, and you can try as many different paths among the links as you have the time to try.

Of course, if you're pressed for time, you probably shouldn't go exploring. But as you have some free time, you might experiment by searching for a topic of interest, not one about which you have to find information. You might let gophers lead you to new information or make different selections than you usually make. You might work with a new database or a different database than you usually work with. As you work with the Internet, you'll discover that new databases, WWW sites, and home pages are constantly being added to the network, all of which give you plenty of new places to explore. And, if you subscribe to a commercial service, you'll have opportunities to learn about additional services and visit new sites.

The best strategy for working with electronic journals, the WWW, and other networked databases of secondary sources of information is to experiment whenever you can. But when you have to locate information under a deadline or for a specific project, develop a search plan of keywords and databases where you can begin your search.

When you work with CD-ROM-based sources, you should have a search plan. Perhaps you know of authors or titles of specific publications that you need to find, because when you purchased or were given access to the disk, you knew that the documents you needed were stored somewhere on that disk. Then you can search by author, title, or publication to find the source you want. If you need to find more information about a subject, you might plan a list of keywords by which you can search the disk for all possible information about that topic.

Summary

Too often, technical communicators become enamored with the technology of communicating and researching. They like the new tools and toys associated with electronic information. But keep in mind that behind the bulletin boards, e-mail systems, and computer screens are people who still value the friendliness, accuracy, and timeliness of good, old-fashioned communication.

In research, the proliferation of sources available through networks or becoming more frequently offered on CD-ROM, for example, means that technical communicators need to develop greater facility in working with electronic, as well as paper and in-person, sources of information. This is especially true with secondary sources of information. As more sources are made available, and because many of them deal with research in progress, technical communicators need to have that immediate access to what is being presented online.

Therefore, if you are interested in a subject, you should find out if there are newsgroups, mailing lists, and discussion groups of people who share that interest. Research in progress, comments about other e-mail messages sent to members of the group, notices about special events and publications, interactive discussions of current topics, and other timely information will be useful to you daily. To keep up with all that's being presented, published, discussed, considered, and so on, in our profession and within your specializations, you should conduct research (and help others with their research) as actively as possible using electronic media.

Research Activities

1. Subscribe to a mailing list of people who share a research-related interest. As you work on a current project, ask your colleagues on the list for information, advice, or criticism.
2. If you need to conduct an informal survey as part of a current project, design an e-mail, as well as a print version. Solicit responses from people through e-mail.
3. Experiment with search strategies using the Internet. How many databases can you find to search? Which subjects are covered in each? What kinds of information can you locate on the WWW?
4. Analyze several home pages that you've found particularly helpful. Then design a home page for you, your company, your department, or your university. What kind of information do you want on your home page? What kinds of information can you share with others on the WWW? How can you collaborate via the WWW?

For Further Reading

Aboba, Bernard. 1993. *The Online User's Encyclopedia: Bulletin Boards and Beyond.* Reading, MA: Addison-Wesley.

Allen, Nancy. 1993. "Learning E-mail Skills." *Technical Communication* (November): 766-68.

Alspach, Ted. 1995. *Internet E-mail Quick Tour.* Chapel Hill, NC: Ventana Press.

Berners-Lee, Tim. 1994. "The World-Wide Web." *Communications of the ACM* (August): 76-82.

Branwyn, Gareth. 1994. *Mosaic Quick Tour for Windows: Accessing and Navigating the Internet's World Wide Web.* Chapel Hill, NC: Ventana Press.

Braun, Eric. 1994. *The Internet Directory.* New York: Fawcett Columbine.

December, John. 1994. *The World Wide Web Unleashed.* Indianapolis: Sams Publishing.

Drummond, Rik. 1994. *LAN Times E-mail Resource Guide.* Berkeley, CA: Osborne McGraw-Hill.

Eager, Bill. 1994. *Using the World Wide Web.* Indianapolis: Que.

Every Student's Guide to the Internet. 1995. San Francisco: McGraw-Hill.

Hahn, Harley, and Rick Stout. 1994. *The Internet Complete Reference.* Berkeley, CA: Osborne McGraw-Hill.

Hayes, Brian. 1994. "The World Wide Web." *American Scientist* (September): 416-20.

John, Nancy. 1994. *The Internet Troubleshooter: Help for the Logged on and Lost.* Chicago: American Library Association.

Lemay, Laura. 1995. *Teach Yourself Web Publishing with HTML in a Week.* Indianapolis: Sams Publishing.

McArthur, Douglas C. 1994. "World Wide Web & HTML." *Dr. Dobb's Journal* (December): 18-26.

Notess, Greg P. 1994. "Lynx to the World-Wide Web." *Online* (July): 78-82.

Schatz, Bruce R. 1994. "NCSA Mosaic and the World Wide Web: Global Hypermedia Protocols for the Internet." *Science* (August 12): 895-901.

Schulzrinne, Henning. 1992. *Voice Communication Across the Internet: A Network Voice Terminal.* Amherst: University of Massachusetts at Amherst, Department of Computer and Information Science.

Tittel, Ed. 1994. *E-mail Essentials.* Boston: AP Professional.

Turlington, Shannon R. 1995. *Walking the World Wide Web: A Listings Guide for Multimedia Resources on the Internet.* Chapel Hill, NC: Ventana Press.

Varhol, Peter. 1995. *E-mail: Achieving Local and Global Communications.* Charleston, SC: Computer Technology Research Corp.

Vetter, Ronald J. 1994. "Mosaic and the World-Wide Web." *Computer* (October): 49-57.

CHAPTER SEVEN

Personal Experience and Knowledge

Research serves two primary purposes. Through research, we discover what we don't know, and we increase our personal knowledge through that act of discovery. Second, through research, we verify, test, and validate what we believe to be true from our other research. As you read in Chapter 2, when you are given a communication assignment, you start by taking an inventory of what you know about the assignment, then determine what you don't know and plan your research requirements.

When you are given a communication assignment, you should focus on what you know about the subject, the available information when you're given the assignment, and the information you know or anticipate that you'll need as you complete the assignment. As you focus on specific information, you'll use it to develop your research and the project. But the result of your work will not be just whatever you discover from your focused research. It will instead be the result of your focused research as it is affected by all your prior experience and knowledge.

The purpose of this chapter is to define personal experience and knowledge, provide some methods for taking an inventory of your experience and knowledge, and suggest some group and individual activities through which you can increase your experience and knowledge.

Defining Personal Experience and Knowledge

Personal experience is simply defined as that which you have done before. Experience suggests involvement in an activity, and certainly the idea of experience makes no distinction between whether you were a willing or an unwilling participant in the activity. Experiences may be general or specific. For example, if you've written something, you have general experience in writing. If you've written a newsletter, you have specific experience in writing newsletters. If you have written newsletters online for electronic dissemination, then you have specific experience writing newsletters for one medium.

Personal knowledge, on the other hand, is that which you have learned from experience, observation, or other research. You can know something without having experienced it, because knowledge can be theoretical. For example, with many scientific theories, researchers assume that something is true, even though they have not experienced it. They may assume that they know how a blue star is created, but they probably have not observed that process. In a more practical example, you may understand how a newsletter is written and produced; thus, you know the physical attributes of a newsletter article and the theory about how an article is written. But you don't have previous experience in writing newsletter articles, and when you attempt to write one you may find that the application is more complex than the theory.

You can also know something because you have experienced or observed it. Experience is frequently used as a marker by which employers or experts gauge knowledge about and skill in a specific field of work. The assumption that experience is equated with knowledge is most clearly expressed in job announcements, in which a particular degree might be required, but a certain number of years of experience are also required. The more experience a person has equates to greater skills and abilities, as well as a better understanding of the job; thus, the need for training is reduced and the speed with which one can become productive is increased. Observation can also play an important part in the knowledge equation. As you've gained experience, you've also observed the way people do their jobs, equipment operates, and processes take place. Although firsthand experience is a greater source of knowledge, you can also gain knowledge by observing what goes on around you.

As professionals, technical communicators are constantly called upon to draw from a range of personal and professional experience and knowledge. How well you succeed may be a matter of how well you are able to recall what you know and apply it to new situations. But, certainly, that which you have learned or done most recently, most often, or most successfully will be that which comes to mind first. As a researcher and professional, you need to develop some activities that help you gain immediate access to your complete range of knowledge and experience.

Taking Inventory of Experience and Skills

If a career counselor told you to take an inventory of your skills and experiences so that you could write a good resume for a job and then interview with confidence, you'd probably understand the logic. You should, in fact, take a personal inventory when you're preparing for a job. However, you should also complete some personal inventories even after you are on the job.

For example, you should periodically take new account of your skills that result from your life experiences. When you do the first, or even the first few inventories, you consider your entire life, including the full range of your activities in school, volunteer groups, and all the jobs you've had. You consider the skills you learned from participating in each activity. Then you chart your inventory in this way:

Activity ➡ Skill ➡ Supporting description of how the activity led to the skill

If you were the newsletter editor for a chapter of a professional association, you have several skills that were developed through your work. You probably had to solicit articles from writers, communicate with writers throughout the revision and publication process, edit the articles, design the newsletter, and work with a printer to produce copies. In addition to communication skills, you may have had to complete a budget, supervise other volunteers, and manage the activities within a time limit. All of these are valuable skills.

These general skills may be further described in more specific details. You wrote e-mail messages, as well as letters, to communicate with writers; therefore, you enhanced your online communication skills and maybe learned how to operate a new application. To design the newsletter, you worked with desktop publishing software, laid out each page, scanned photographs into the software, selected clip art, and wrote captions and headlines. Each general skill—such as communication, management, or design—can be detailed. These activities should be described in your personal inventory.

You can inventory your knowledge in the same manner. Initially, you might think about what you know in terms of classes or workshops you've taken, presentations you've heard, or literature you've read. But you also learn from your experiences. Thus, you create a chart similar to the following:

Activity ➡ Knowledge ➡ Supporting description of theoretical knowledge

For example, if you observed a lunar eclipse, you have a good idea of how and why the eclipse takes place, even if you've never taken an astronomy class. You also have a better idea of how people react during an eclipse, and why, historically, eclipses were accorded magical power over people. If you helped a friend develop photographs in a darkroom, you have gained knowledge about the photographic processes that you helped complete. Your knowledge has been increased about the chemicals needed to develop photographs, the time frame for the process, and the specific steps. Your activities helped you develop new knowledge and frame your new knowledge within what you already knew about this subject.

Initial inventories may provide you with only a one-to-one relationship. For example, you may equate accurately that you wrote a couple of newsletter articles, so you have some experience in writing newsletter articles. However, as you review and continue to work with your inventories, you might look beyond the one-to-one equation. As you prepared for writing that newsletter article, for example, you may have done some research about newsletter articles. Perhaps you read about line lengths, use of graphics, audiences who read newsletters, and typefaces, but the only information you used was that which you discovered about audiences. What you did provided you a skill, but the skill does not indicate the full range of your knowledge.

As one who conducts research either to gain theoretical knowledge or to apply the results of that research, your knowledge and experience, as well as that of any others who may be working with you, come to bear upon the methods you use and the results you achieve, more so, perhaps, than is easily definable in any particular situation. Therefore, the more you know about your range of skills and knowledge, the more efficiently and confidently you consider new projects.

In every situation, you work from a personal perception and with ideas and attitudes that are the results of others' perceptions. How you perceive information and tasks influences the work process, sometimes overtly and sometimes subtly. What causes your perceptions is the cumulation of your experiences and knowledge at any given time. Perception may not affect the tasks, time lines, or budget, but it affects the attitude with which you approach work and co-workers.

Perception is, however, not always the result of complete and accurate information. As we can misunderstand the words and actions of others, we can also misperceive ourselves. You can misunderstand information you have because of faulty perception, and you can also misjudge your skills and knowledge. The inventories you complete help provide you a clearer understanding of your knowledge and capabilities.

On the other hand, those inventories provide you information about what you can do and what you know, but they do not necessarily provide you information about why you do what you do, what energizes or enervates you, and what affects the way you perceive the tasks you need to accomplish and the people who work with you. To understand yourself a little better, you also complete inventories of your personal and professional goals and objectives, of what you like and dislike about what you do. For example, you may discover that the reason you're dissatisfied with your job is because all assignments have to be done in a hurry and you don't feel comfortable writing without ample time to conduct background research. For you, completing enough research before you begin writing is almost an ethical issue. The documents you produce don't provide the information as well as they could. Therefore, when you take an inventory of your work objectives and your personal goals, you discover that one of your objectives is to find more time to spend on the documents you write. Your personal goal may be to be a craftsperson with a well-known reputation for thoroughness. If your current job doesn't provide you with ways to meet your objective and work toward your goal, you might begin looking for jobs or assignments that will let you fulfill your personal needs.

Knowing why you do or don't like something may create a cause-and-effect relationship. Knowing about your likes and dislikes can provide the impetus to change. You might seriously consider changing jobs, or you might review current procedures and recommend changes that would alter the way you work. For example, you might attend, and also suggest to your department that everyone attend, workshops or training seminars on time management.

Knowledge about your preferences can also help you manage projects and people. You design time-task breakdowns to provide more time for work in those areas about which you feel less confident. As you develop a schedule for others, you remember their strengths and weaknesses, likes and dislikes. You then develop a schedule to help them work with their strengths and likes, whenever possible. You determine who is better doing certain tasks and who works well together. Knowledge of others', as well as your own preferences and practices helps you become a better manager.

For example, Nancy is a technical writer who has been given the job of writing a software manual. The software is still being developed, so Nancy must first rely on her previous experience with other software, writing, and manuals, as well as her memory of her first experience working with a computer. Nancy created a few lists of relevant skills and educational experiences to help her remember all the personal experience and knowledge she might bring to this project (see Figure 7.1).

Nancy knows what should go into a computer software manual. She's written a few, and her degree is in technical writing, so she knows what ideally should go into a good manual. She remembers vividly not only her pleasant experiences with computer work, and her years of experience using computers, but she also knows what it's like to lose information, crash a system, and otherwise have a bad day working with computers. She can use her experience and knowledge very practically to help her determine what should go into the manual and to provide her with a starting place for outlining the manual's sections, perhaps for a doc plan she'll create. But just as important, she can empathize with her audience of novice computer users who rely on her for accurate information and an interesting, guiding approach to their work with new software. Because Nancy understands how her audience feels and what they may fear, enjoy, and need to know, she can more easily provide them that information in a usable format.

General inventories of your strengths and weaknesses, therefore, are excellent personal research tools. They provide information that you can use in a variety of circumstances. By understanding the relationships among experiences, knowledge, and perception, you better understand

Experience	Knowledge
Work with WordPerfect, Word, Wordstar, Lotus, Excel, Pagemaker, dBASE, Quicken, FoxPro, AmiPro.	Computer training classes
	Degree in technical writing
Wrote two manuals last year.	
Trained 25 people in the company to use our customized database.	
More than eight years of experience working with computers in some way.	
Remember the first time I "blew up" a system.	
Remember the first time I didn't back up a day's work and the system crashed just before 5:00.	
Just completed an online documentation project.	
Designed icons for online Help.	

Figure 7.1 A list of relevant experience and knowledge.

at least some reasons audiences respond so differently to the same kinds of information.

Brainstorming in Groups

Brainstorming individually, using either your computer or just a sheet of paper to record your ideas, is an effective way of determining what you know about a subject. But brainstorming as an individual works best when you are the solitary author or when you are considering a topic for a project about which you may have sole or primary responsibility. Sometimes, however, you may be pursuing a project as part of a group. The entire group will be involved in all or parts of the decision-making processes, including focusing the ideas the group needs to pursue further.

Brainstorming as a group allows everyone to present his or her ideas. It is particularly effective because ideas come from a variety of people with a range of experiences and knowledge, and therefore a variety of perceptions about the information.

Group brainstorming can, and frequently does occur, especially during initial meetings of groups beginning new projects. These impromptu group brainstorming sessions can be excellent sources of information, but they can lack the focus to make them most effective. Group brainstorming sessions are most productive when there is some degree of planning. This planning may involve scheduling meeting times and places, creating an appropriate atmosphere for idea exchange during the meeting, and recording and following up effectively.

The atmosphere for a group meeting results from a number of factors, including the reason for the meeting, the timing of the meeting, and the method in which the meeting is conducted. When you go to meetings, you want them to be productive; to have a specific, attainable objective; and to be scheduled when they provide the least interruption to your required daily activities. Part of planning a group brainstorming session is conducting the meeting so that it creates an inviting atmosphere. Everyone should be comfortable, able to see and hear everyone, and be able to be seen and heard by everyone else. Distractions should be minimized and discussion encouraged. Flipcharts, marker boards, notepads, laptops, or other idea-capturing tools should be plentiful. There should be plenty of time for discussion, but not so much time that the meeting becomes boring and unproductive.

Planning further involves ensuring that all participants enter the group session with the same kinds of information. For example, if the session is the first step in planning after a project has been assigned, then all group members should have the same copy of the assignment. If the group needs to do any reading or other research prior to the session, then the research responsibilities should be clearly outlined and provided so that each participant has the time to do the research required.

Brainstorming sessions should be recorded in progress. There are a number of different ways to record such a session. It can be video- or audiotaped, but such recordings lend themselves to two distinct problems. First, some people don't respond freely as they think of ideas when they know they are being recorded, so they filter their ideas before presenting them. Second, taping also requires that someone review the tape, screen what will go into a file of ideas to be presented to the group, and record those. Again, the result of the brainstorming session may be incomplete because any transcription is influenced by the transcriber's editorial judgment.

The session may be recorded on computer, with one person writing the ideas as they occur. Recording in this manner allows everyone to get the same results, which can be provided electronically or on a printout. Recording ideas on computer during the brainstorming session shields them from the participants. Thus, only what the participants hear and their own thinking generate new ideas. However, it also can lead to some people thinking through their ideas to consider whether they're repeating what's already been stated.

Recording ideas on a marker board or another surface is yet another approach to recording ideas as they are presented. All ideas are constantly visible, so people not only hear them but can also read them. Seeing and hearing together is frequently an effective prompt for people, especially because people tend to review the list as they think and the reviews also prompt new ideas.

The way a session is recorded will affect the way the session is conducted. But brainstorming also requires that all members of the group feel free to respond without their ideas being subject to question or comment by another group member. Certainly, getting everyone to respond is a matter of group dynamics and individual personality, but it can also be a result of planning and conducting the group session.

Brainstorming sessions must follow general rules of etiquette in that everyone must feel that he or she can participate without his or her ideas becoming the immediate subject of comment and evaluation. During a brainstorming session, all ideas are relevant, even if they may be modified or discarded when the ideas are sorted.

Brainstorming sessions are designed to provide the opportunity for the unedited, free flow of ideas, but they should be directed toward some goal or objective. Thus, the leader of the group must clearly define the objective for the session, the parameters that will focus people's thinking onto ideas that can be developed to achieve a larger objective. If, for example, the group is required to produce a corporate annual report, then the first question might be as general as What have we done this year? Other questions to help keep the group focused on the company's achievements might be What are we proudest of? What problems did we have? How did we overcome them? What do we have to put in an annual report? Where are we headed in the coming year? Later questions might focus on the presentation of the information for the report, once the group has generated several ideas about the report's content: What's a colorful

Plan
- A clear objective
- A convenient time to meet
- A regular meeting time that doesn't interrupt other scheduled work activities
- A single agenda item
- A comfortable, distraction-free environment
- An efficient method of recording the results of the brainstorming session
- A clear process to review the ideas (deleting, adding, sorting, classifying)

Figure 7.2 Planning a group brainstorming session.

way to present statistical information? How many photos do we want? What would make a good cover? What's new and different in other companies' reports? How can we capture our company's spirit in a different way? Figure 7.2 outlines some considerations when you lead a group brainstorming session.

Clustering

Clustering is another method of thinking about what you want or need to write. It is similar to freethinking. However, clustering is more effective when you have a little more experience with writing and know the kinds of subject areas you'll need to include in a project, even though you may need to conduct more research about the specific subject. Although clustering to generate ideas starts with the presumption that you are working in the abstract—that is, with only a general subject area to guide you—clustering to recall experience or knowledge starts from some specific categories of information.

Clustering allows you to jot down groups of items that you know or groups of questions about what you need to learn. For example, Figure 7.3 shows a clustering of subject headings.

As you either discover a new heading you want to consider, or find information you wish to put under each of these subject headings, you can add that information. Clustering is similar to brainstorming, with one important difference: When you get an idea or think of an example, you place it near the word or phrase that prompted the new idea or example. You cluster the ideas that are generated around the word or phrase you were considering when you thought of another word or example. Clustering exercises should take no longer than four or five minutes, and

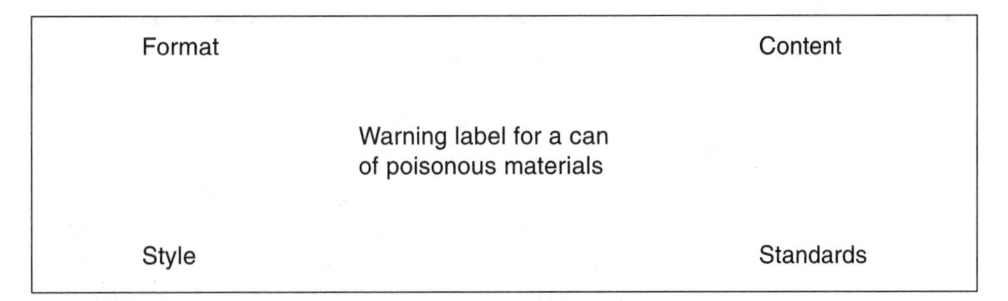

Figure 7.3 Clustering plan for a writing assignment.

during that time, you should concentrate only on the words or phrases used in the exercise. When you've finished, you have several groupings of information clustering around your primary subject heading, as in Figure 7.4.

In this clustering exercise, the writer even created some phrases that might be used in the hazard information itself. Although the warning is still far from being written, the writer tapped into his memory of other warnings he and others have written and jotted down some ideas about how those warnings might apply to the language needed in this warning. Of course, there are still a number of questions that need to be answered and further research that needs to be done.

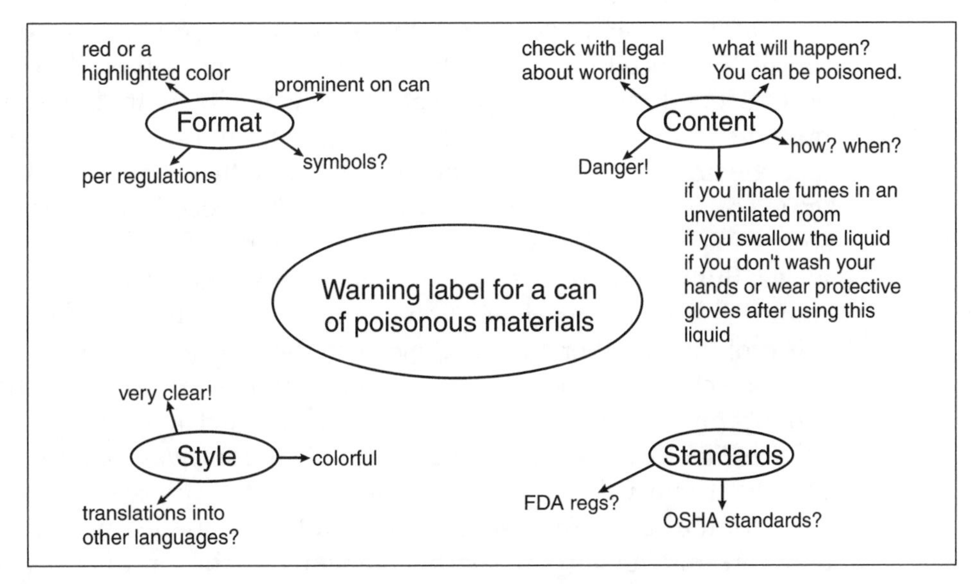

Figure 7.4 Clustering for a writing assignment.

Gaining More Personal Experience

By participating in society, working with others in a company or within a university or college, reading professional journals, observing the world around you, and just living, you're gaining valuable personal experience that you can use in your writing. Without concentrating on it, you're conducting research about the way people react to situations, how the world operates, and what is currently popular—just by doing the things you typically do every day.

Although this general experience may be useful when you write, because it helps you understand the people who use your information and become aware of what currently is important to them, you sometimes need more specific experiences that you can apply to a project. For example, Nancy, the technical writer assigned to write a software manual, has a great deal of experience and knowledge she can already apply to her work. However, because the software is still being developed, she'll need additional experience working with it so that she can accurately provide information for her audience. Of course, Nancy probably will use other research methods, too, such as talking with subject matter experts (SMEs) and perhaps assisting with the product's usability testing. But personal experience working with the product is a valuable source of primary information.

If at all possible, you should work with the product or observe the process or procedure about which you're writing. Although past experience is good, specific new experience working with the subject can provide you with detailed information and an attitude or approach to take toward the subject. Of course, talking with the product developers or personnel who perform the process or the procedure is another good source of information, as you'll read in Chapter 8, Informational Interviews. But getting hands-on experience provides you with personal insight—as well as experience you can use as a source of information for this project and other similar projects in the future.

For technical writers like Nancy, the experience of working with a new product provides time to troubleshoot the product and work out any bugs or investigate differences between the way the software should work and what it currently does. It gives her the chance to explore new features, get a feel for the benefits of using the new software, and learn the differences between this version and earlier versions of the product. By learning directly what the software can and can't do, how it operates, and what it looks like, Nancy can write more precisely. She's gathered a great deal of information in a short time, and she can mentally—and perhaps in writing—compare this new product with her experiences with other software.

Many technical communicators enjoy their work primarily because they can develop new skills, learn by observing people and machinery at work, and use cutting-edge equipment and procedures just being developed. One freelance technical writer, Lynn, for example, has learned about a variety of different subjects, just by conducting research that provides her with a wealth of personal experience. In just the past 10 years, she has learned how a printing press operates and how to set up and tear down a press, learned to use commercial and customized software, observed the production of railroad cars, observed a surgical procedure, visited a hazardous site that was being remediated, and toured a robotics shop. The information she gathered through working with equipment, observing people doing their jobs, observing equipment in operation, taking apart and putting together components, and visiting field sites helped her write training materials, safety manuals, reports, and proposals.

As you gather information about a product, a process or procedure, or even a place, you might keep in mind questions found in Figure 7.5.

Who

- operates the equipment?
- performs each task?
- makes decisions about this process or procedure?
- can make changes to this process or procedure?
- can make changes to the equipment?
- designed the equipment?
- manufactured the equipment?

What

- occurs first, second, third, and so on?
- occurs simultaneously?
- does each component (e.g., key, lever, knob) do?
- does each person working with this process do?
- is supposed to happen?
- actually happens?
- is the operator supposed to do?
- does the operator typically do?
- should the operator do in an emergency?

(continues)

Figure 7.5 Guiding questions during an observation or hands-on work.

does the operator see at each step of the process?
does the operator hear at each step of the process?
does the operator smell at each step of the process?
does the operator taste at each step of the process?
does the operator feel, touch, maneuver, etc., at each step of the process?
should the operator wear?
is an acceptable degree of variation in the procedure?
does the operator do if something doesn't work properly?
is the name of the procedure?
is the name of the product?
is the name of the equipment?

When

should each task take place?
should each task not take place?
does an alarm sound or a warning notice appear?
does instructional information (e.g., an on-screen prompt) appear?
does a task have to be completed?
is it acceptable for a task to be completed?
is it too late to complete a task?

Where

are the materials needed to complete the task?
is the equipment needed to complete the task?
is each component or subassembly needed to complete the task?
is nearby safety equipment?
is information about the product or process?
are other people during this process or procedure?
is the process usually completed?
should the process ideally be completed?
should the process never be completed?
can replacement parts be ordered?
can repairs be made?
can new equipment be installed?
can enhancements or changes be made?
should changes never be made?

Why

is this the best way to complete the task?
does the equipment operate this way?
do people feel as they do about the product?
is this product better than an earlier version?
is this product better than competitors' products?

Figure 7.5 Guiding questions during an observation or hands-on work. (*Continued*)

As you gain hands-on experience, observe equipment in operation, observe people at their jobs, or visit a site, you should document your findings as meticulously as if you were conducting an experiment. In many ways you are—you're testing what happens if you vary a process, press a key, turn a lever, modify the equipment, and so on. You should take notes about what you see, hear, smell, taste, touch, and feel, as well as what you observe about others and the equipment they work with. You need to be very descriptive, so that when you sit down to write, you'll remember vividly what occurred and you can include accurate details in even a first draft of your document.

Summary

Although the place where each of us begins research may be different, we each have unique experiences and knowledge from which we can start. Through freewriting, group brainstorming, clustering, or other ways of capturing information, you can get a good sense of what you know, what you've done that can be applied to your project, and what else you need to know and do.

One exciting element of working as a technical communicator is the variety of subjects about which you might need to learn, just to keep up with your area of science or technology and the changes in the way people communicate. Technical communication is a very dynamic profession, and your research will often involve working with a product, learning to do a new job, developing a new skill, and observing the way people and equipment interact as the workforce changes. Your experience and knowledge are valuable assets in your work as a technical communicator, and you should seek those situations that allow you to gather new information and develop new skills. Because technical communication changes rapidly, you must keep developing a repertoire of experiences and keep adding to your knowledge base. In turn, what you know and have done helps you to learn and do more, and the cycle continues.

Research Activities

1. Make a list of your knowledge and experience as they apply to a project you're just beginning or plan to begin. Then make a list of

new skills or knowledge you'll need to develop during the course of the project.

2. Visit a lab or a field site and observe what goes on. Practice taking notes about a process or procedure, either mechanical or human, you've observed.
3. Gain a new technical skill. You might participate in a workshop, take a course, follow instructions, or otherwise learn how to do something. Then practice completing this activity until you are comfortable with your new skill.

For Further Reading

Aiken, Milam, Jay Krosp, and Ashraf Shirani. 1993. "Electronic Brainstorming in Small and Large Groups." *Information Management* (September): 141-48.

Clandinin, D. Jean, and F. Michael Connelly. 1994. "26 Personal Experience Methods." In *Handbook of Qualitative Research*, eds Norman K. Denzin, and Yvonna S. Lincoln.Thousand Oaks, CA: Sage.

LaPlante, Alice. 1993. "'90s Style Brainstorming." *Forbes* (October 25): 44-61.

LeCompte, Margaret Diane, Judith Preissle, and Renata Tesch. 1993. *Ethnography and Qualitative Design in Educational Research.* 2nd ed. San Diego: Academic Press.

Mason, Nondita. 1994. *Writers' Roles: Enactments of the Process.* Fort Worth, TX: Harcourt Brace College Publishers.

CHAPTER EIGHT

Informational Interviews

Without question, printed, audio, audiovisual, and other sources of produced information are valuable sources for technical communicators. From these sources, you get historical information about various subjects. You can compare sources for completeness and reliability. You can read somewhat at your own pace (considering, of course, deadlines and other obligations that affect time). With so many possibilities for electronic access to various libraries, you also have great access to published sources. However, these published sources in some instances just aren't available, or appropriate, for getting the information you need. When they aren't, you consider conducting informational interviews. Generally, informational interviews can be defined as one-on-one interviews in which one person seeks information orally from another.

For technical communicators, informational interviews are a primary method of gathering information in a number of instances, for example, the following:

- To present themselves to companies or groups with whom they'd like to work.
- To get information that isn't produced in some other form.
- To verify information they discovered from some other form.
- To update information from published sources.
- To clarify information.

To use informational interviews most effectively in your search for information, you carefully develop a strategy to conduct that interview. You should know how to research for, prepare for, conduct, and follow up an informational interview. This four-step plan for a good informational interview helps you consistently and methodically conduct your research with SMEs.

Much technical or scientific information can be gathered only from the people who are busy developing new products, updating procedures, and adding new information to what is commonly known about a subject. The only way to gather information about such fast-changing and cutting-edge topics is to interview the SMEs who know about them.

However, gathering information for assignments is not the only use for an informational interview. It is also an effective strategy to use during a job search. In this chapter, you'll read more about planning and conducting informational interviews for job searches and assignments.

Informational Interviews as Job-search Tools

If you're a new technical communicator, or an experienced technical communicator who's interested in a career change, conducting informational interviews is probably an important part of your job-search strategy. Searching for a first or a new job or changing jobs is more than a process of checking the many sources that post job announcements and presenting your credentials to those employers who appear most interesting. It's a matter of selecting a job that's compatible with your skills, your expectations for current work, your expectations for your career, and your values and principles as a technical communicator. Selecting a place to work is a process of matching both personal and professional interests. It's difficult, at best, to determine if these matches exist just from reading job solicitations.

Whether you are a new or an experienced technical communicator, you may find yourself in a situation in which you are either thinking about getting a job or considering changing jobs. If you're a freelance technical communicator, this process of locating potential new jobs is a regular part of your work. In these situations, the informational interview can be an excellent tool to help you decide whether you want to change jobs or solicit a position from a particular company.

Also, there are times when you may find yourself job hunting, and no solicitation notices appear to match your interests and needs. On these occasions, you should consider the informational interview as a job-searching tool.

Informational Interviews for On-the-Job Information Gathering

Technical communicators often gather information from SMEs as one of the most important primary sources of information. If you're working within a company, you usually know, or are told, who are the product designers, programmers, engineers, scientists, and so on who are developing information or a product. It becomes your job to learn from these experts about a subject. These people are often the only ones who can provide you with accurate information about the subject, because they are creating a new product or finding new information that has not been printed or otherwise distributed. The only way you can get the information you need is to talk with the experts.

Sometimes you have to talk with several experts, each with specific information that no one else has. These people, and you, all have different schedules and a limited amount of time to talk. In your daily work, you may have to juggle several interviews within a few days. That's why it's especially important to have an effective approach when you have to schedule several interviews and keep track of a lot of information from different experts for one project.

A Four-Step Plan for Conducting an Informational Interview

Each time you prepare for an interview, you need a strategy that lets you make sure you've done everything necessary for a good interview, before, during, and after the actual interview. That strategy becomes especially important when you're working on several projects at once and/or you have several people to interview for the same project. When you prepare for an interview, you do these tasks:

- Research
- Prepare
- Interview
- Follow up

Step 1. Research

Before you schedule an interview, you conduct the background research necessary for the interview. If your employer or client didn't provide you with a list of SMEs to interview, your research can provide you with names of people who might be good sources of information. Secondary sources may help you learn about some SMEs' credentials, or you might need to talk with people who work in your company to learn with whom you should conduct interviews.

Research for Background Information

If your project involves subjects outside your area of expertise, you probably have to learn some basics about the subject before you feel comfortable talking with an expert. You might need to brush up on your information and update your knowledge. Even if you're familiar with a topic, your background research might uncover new information about which you were unaware.

Even though your area of technical or scientific expertise may be the area about which you'll be writing, you still may need to brush up on the latest information. Part of the challenge of being a technical communicator is keeping up with daily changes in products, services, and the general knowledge base within the numerous technical and scientific subject areas in this profession. Before you begin asking questions of your experts, you may need to read background information, get some experience working with a product, see the product in operation, observe the process, or otherwise brush up on your information.

Sometimes you're asked to write about subjects for which you have little prior information. In these cases, you may need to read a number of good secondary sources to understand the basic terminology and concepts of the subject. You need to know enough to determine what you have to ask experts.

When you conduct background research about a subject, you ask yourself these types of questions:

- What has changed recently in this subject area?
- What is the most important information currently available?
- How can I update my information?
- What are new terms that have been added to experts' vocabulary?
- What products or processes are similar to the one I'll write about?

Research about a Company

If you're searching for a job, you might need to learn more about the company, its services or products, and the type of job you might be interested in. When you research background information for a job search, you plan to interview only those persons in companies where you have a real interest in working. If you believe you're interested in a company, find out all you can about that company.

You can get information about most companies through a variety of sources. If you're near a college or university, for example, you might check with the placement office or employment service, especially if you're an alumnus of that institution. Placement services have different kinds of literature about companies. University, city, or community libraries also have varieties of sources from which you can get information about companies. A look through the reference section of a local library might show you references such as Standard and Poors, Fortune 500 and 1000 guides, business directories, employment outlook references, and other sources that provide information about companies.

If the company is local, you might set up a visit or a tour, during which you probably will be rewarded with brochures, in-house newsletters, and other kinds of promotional literature that discuss the company, its services, and its products. If the company isn't within driving distance, a phone call can often help you get corporate literature.

Two other sources of information about companies are professional organizations and electronic sources. As one involved in technical communication, either as a student or as a working professional, you should consider it a professional obligation to participate in organizations that represent technical communication. One of the benefits accruing to you from participation in such organizations is the opportunity to meet people who represent companies in which you may have some interest. Particularly if the company you're interested in is local, chances are that you can meet representatives of that company in local professional meetings. However, when the companies are not local, you still have the opportunity to learn much about the companies through your work in professional organizations.

For example, as a member of the Society for Technical Communication (STC), you can contact chapters in the area in which you are interested in finding employment, and chances are strong you can get local employment information. Also, international, regional, and local meetings and conferences attract people from a variety of industries. Most of these people probably share to some extent your professional interests and are generally quite amenable to discussing their work and their industries. Meeting and talking with people informally are excellent ways to gather research about companies and make contacts with people who are employed by the companies in which you're interested. At the same time, you are gathering information about what companies are doing and what they anticipate in the future. These kinds of information can be very helpful in suggesting new areas of research you might want to pursue.

Electronic sources also provide a great deal of information about companies. To attempt to provide in this book the various kinds of electronic information you can find about companies would be essentially to provide you outdated information, for the information available on computer changes daily. The kinds of information you want to solicit about a company are the best guide to the kind of searching you want to do electronically. For example, if you're interested in finding whether a company has been grant funded for its research, or if some employees' positions are grant funded, you can electronically search various federal and state granting agencies to see which grants have been awarded to the company.

Once you've learned what an organization does, what appears to be its primary aims and objectives, and its philosophy as an organization, you can begin to think in terms of whom you should contact for an informational interview. Many of the same sources, for example, STC and other professional organizations, may help you decide who you should contact for an interview.

Selecting People to Interview for a Specific Assignment

When you plan an informational interview as part of your regular job, the information may lead to a product, such as a document. You may know about the company, because you're a current employee or you're a freelancer who has worked previously with the employer. But you may need to determine who will be the best people to interview. Your background research provided you with names of experts who can provide you information for your project. The SMEs may be within your company or perhaps even nationally renowned. The better known the SME, the more difficult it may be for you to get an interview with that person. Your selection of SMEs should be large enough so that if you can't get an interview with one person, you have others to call upon for an interview.

Your background research about the person or people you'll interview may be simplified if you're conducting in-house interviews. Probably you've interviewed the experts who will provide you information about a project, because you've interviewed them before for similar projects. Even if you will be interviewing new people, you may be directed to the people to interview by your manager or supervisor.

If you need to determine who will be good people to interview, answer these questions:

- Who are experts in this field?
- Who specializes in this subject?
- What are this person's credentials?
- Who can offer me a different perspective on this subject?
- Who can I interview within the time frame needed for this project?
- Who is it feasible to interview?

When you've identified the best people to provide information, next make sure that you can secure an interview with each of them. Time, distance, and money may be some obstacles to getting the ideal person to interview, but you should be able to locate a number of experts who are available and will agree to be interviewed.

The research you conducted that helped you decide where you wanted to interview for a possible job may also provide you with the names of people whom you can contact. When that happens, you have a start. When it doesn't, then you need to discover specifically with whom you wish to talk. One method of doing this is to call the company directly, tell whoever answers (such as a receptionist or a switchboard operator) your interests and ask him or her to give you the name of the person or persons who work in the areas in which you are interested.

When you have the name of a person, contact him or her. Again, you have a choice of making initial contact by telephone, by letter, or even by e-mail. The keys to making an initial, effective contact are the same, and they are listed in Figure 8.1.

In a phone call, the conversation may not allow you to note without discussion all of these kinds of items, so remain flexible in the course of the discussion.

If you're writing a series of e-mail messages, your first message may introduce yourself and your interests, indicate the need for and the benefits of an informational interview, and suggest a possible convenient

How to Conduct an Informational Interview about a Job

- Introduce yourself and your interests.
- Indicate what prompted your call, letter, or e-mail message specifically to that company and person.
- Indicate that you're interested in conducting an informational interview.
- Suggest benefits to the person, or the company, of conducting the interview with you.
- Indicate your flexibility in scheduling an interview, but be prepared with some specific suggestions for dates, times, and places.

Figure 8.1 Conducting an informational interview about a job.

time or method of completing the interview. Later e-mail messages may be used to verify the time and place for the interview, or even to conduct the interview.

Writing a letter is the most formal method of making contact. A letter requesting an informational interview might look something like the one in Figure 8.2.

A letter like this one might be useful in making an initial contact, but writing letters and waiting for responses is a slow process. If you need to contact a potential interviewee through the mail, you should be certain that you have lots of time to negotiate setting up the interview.

Researching the Interview Situation

Research requires more than some basic background information about the interviewee and subject. You also find answers to these types of questions:

How should I take notes or record the information I receive during the interview? Using a traditional notepad, a tape recorder, or an audio- or videotape are all possible ways to take notes. Taking notes using a combination of these methods is preferable. Using a notepad, for example, in combination with another medium allows you to record impressions and new questions or responses. Be aware, however, that using multiple methods of note taking is not always possible. If you'll be conducting the interview in a noisy environment, recording equipment would be ineffective, in which case old-fashioned note taking with paper and pencil is the best method of gathering information. If you'll be

English Department
The University of Findlay
1000 N. Main Street
Findlay, OH 45840

February 1, 1995

Bartley A. Porter,
Communication Specialist
Sprint
P.O. Box 3555
Mansfield, OH 44907

Dear Mr. Porter:

One of my students majoring in technical communication, Janet Davis, is interested in conducting an informational interview with you. She's specifically interested in working as a corporate public communications writer when she's graduated. I heard you talk about your job at an Association for Business Communication meeting, and I think you could provide Janet with a unique perspective about the telecommunications industry.

I'd like to invite you to The University of Findlay so that Janet could conduct an informational interview with you. The interview should take no longer than one hour. After the interview, I'd like to invite you to be my guest at lunch at the University and to take a tour of the technical communication classrooms. Of course, I'd like to schedule the interview on a day when you'll be conducting business in the Findlay area.

I'm specifically interested in your providing the following kinds of information:

1. Credentials students might need to work for your company

2. Internship possibilities within Sprint

Certainly, I will appreciate any other information you can provide me or my students.

I'll call you next week to see if we can schedule a convenient time for the interview. If you would like to contact me in the interim, please call me at 424-XXXX.

Thank you for considering this request for an informational interview.

Sincerely,

Lynnette R. Porter
Associate Professor, Technical Communication

Figure 8.2 A letter requesting an informational interview.

gathering information about a process, such as the way someone does a job, you may be able to videotape or film the process, to make sure that you have all the information in the right sequence. Seeing is better than having to see and then write, rather quickly, all that's going on. Whatever the situation and the interviewee's preference, research what will be the best way of recording the information you're seeing and hearing.

What do I need to obtain before the interview? Your research might indicate the need for security clearances to interview someone in another company or at a government installation. You may need a parking permit, a visitor badge, approval to take photographs or use recording equipment, or other routine permits and approvals before you conduct the interview. The earlier you know that you'll need these documents, the earlier you can make arrangements to get them.

What do I need to arrange before the interview? If you need to travel to a site to conduct an interview, plan your travel arrangements and determine what you have to do to get accommodations, tickets, and so on.

If the SME is coming to your office or your company for the interview, make the environment comfortable and hospitable. You might have water, coffee, or other beverages available. You might make sure someone is available to help the expert find the place for the interview. You might reserve a meeting or a conference room.

What should I wear? If you're conducting an interview in a formal business setting, wear the appropriate business attire for that setting. But technical communicators often interview people where they work, which might be very different locations from an office. You might interview someone on an assembly line, in a warehouse, on a construction site, at an archeological dig, on a boat or in an airplane, in a simulator, and so on. Before you leave for the interview, make sure that you know the appropriate dress. If you need to obtain special clothing or protective devices, such as a hard hat, goggles, or environmental protective gear, make arrangements to get them *before* you go to the interview site. Your research can help you determine what is appropriate for your interview.

Step 2. Prepare

During preparation for the interview, you act on all the research you completed during the planning phase. You schedule the interview with the interviewee, so you know where and when the interview will take place and how you'll record information.

You make the necessary travel arrangements if you're traveling to another site to conduct the interview. You get the equipment or materials you'll need during the interview. You apply for clearances, parking permits, or other documents you need to be at a site. Finally, but probably most important, you develop a set of questions you'll ask during the interview.

Scheduling the Interview

In some situations, SMEs are doing you a favor by taking time to answer your questions. In other situations, it might be part of their job responsibility. In either case, as the person seeking the interview, the responsibility for planning is yours. When you schedule an interview, you should know if you'll probably only need to talk to that person once, or if you should set up a series of shorter interviews.

When you set up the interview, you already have in mind where the interview will take place, how much time will be required, and within what time frame you need the information. The rest of the scheduling is a matter of negotiating a mutually agreeable time for you and the interviewee to get together.

Making Arrangements

Now that you know what you're required to bring to the interview, where it will take place, and what you need to do in the intervening time, you can make the appropriate arrangements for facilities, accommodations, travel, and documents you'll need before and during the interview.

Writing a List of Questions

You should always go to an interview with a list of questions that you know you'll need to have answered. This list will help you keep the interview on track and ensure that you get the basic information you're looking for. The questions should be both short answer and open-ended, but you should always be prepared to adjust the number and kinds of questions you ask in response to how the interviewee responds to the questions. By preparing open-ended questions (how, who, what, when, and why questions), you can get an interviewee to provide more information than just yes or no.

As you write questions, arrange them in a logical order, moving from general, background, and "warm-up" questions to more specific questions. Although you may have hundreds of questions in mind, write the most important questions and revise them until they're clear and direct. Your list may have 10 to 20 questions when you're finished, not hundreds, but if the mix of questions requiring short and detailed answers is appropriate, a few questions can result in more information than several. During the interview, you want to get the information you need, but you don't want to have to fire questions rapidly at your interviewee. Try to fit an appropriate number of questions into the time frame for your interview.

Once you've developed your list, part of your preparation involves becoming familiar with the questions. Know what you want to ask and become familiar with your list so you can ask the questions naturally during the interview, and not leave out any important questions. Note when one response answers more than one question.

Practicing the Interview

Especially if you're a new interviewer, or if you're anxious about the interview situation, as in an informational interview during a job search, you may want to practice conducting the interview. You might have someone pretend to be the interviewee and respond to your questions or ask you questions. You can develop the poise needed during any interview situation.

When you conduct an informational interview in preparation for applying for a job, you should remember that your ultimate follow-up for that interview will be an application. Chances are strong that the person or people whom you interview will also have read your credentials, such as a resume or a vita, and samples from your professional portfolio before the interview. But they'll also review your credentials after the interview. Thus, when they read your credentials, they'll be reading paper, but they'll be remembering you.

If you're interviewing someone to gain general information, not specifically for a job you may want, you still need to be aware of how the interviewee perceives you. What interviewees remember will be the impression they perceive, not necessarily the impression you wanted to present. Never underestimate the power of perception, and always remember that spoken language and body language both create a perception.

As you practice your interviewing skills, you should learn if you have habits that you may not be aware of but that may be distracting to the people you interview. You also need to practice developing a calm, professional, capable demeanor.

Thus, when you interview, you should project yourself as knowledgeable, capable, and confident. The best way to present this attitude is to actually have the attitude, and the best way to have the attitude is to be well prepared.

Step 3. Conduct the Interview

You've already done a great deal of work, and you're now ready to conduct the interview. Most work involved in interviewing takes place before the interview. When you conduct the interview, you concentrate on making the interviewee feel at ease, listening carefully to what is said, being aware of what you're seeing (including body language), and making sure you're recording the information so you can use it later.

Making the Interviewee Feel at Ease

If you work with the person to be interviewed, you already have some kind of rapport with that person, and the interview may be very informal. However, if the interviewee does not know you, you must work even harder to establish your credentials, convince the person to agree to the interview, and establish a sense of friendly professionalism at the start of the interview.

The interview itself must have a professional tone, and usually the tone is formal. You should be the one to offer to shake hands, and you should put the interviewee at ease with some general comments about the site or the interview. Rapport with the interviewee should be established in the first minute of the interview; first impressions do count, and if you have planned for the interview and sincerely are interested in the interviewee, you make the best impression.

You put the interviewee at ease before the interview starts. Getting coffee or taking a moment to chat can make the interviewee feel more comfortable and thus more likely to provide more information. Although you want to keep the interview on track, a few moments of conversation and "getting to know you" can make the difference in the success of your interview.

Paying Attention During the Interview

When you ask a question, listen attentively to the answer. Often, you may be tempted to concentrate on taking notes or otherwise recording the information. But a good interviewer is aware of everything going on in the interview.

Listen to the expert's voice as he or she answers each question. Hesitance, pride, uncertainty, confidence, frustration, pleasure, anger, and other positive or negative emotions can be heard, sometimes subtly, in the interviewee's voice. Therefore, you need good listening and observation skills. Listening to the interviewee is important, because you may think of new questions based on something the expert just said. Tone, expression, and phrasing should catch your attention. An observant researcher can learn a lot by picking out details about the site and the interviewee; the way the interviewee pauses before answering a tricky question, or the way the interviewee completes a typical job task can provide you with as much accurate information about the subject as a formally prepared statement or carefully practiced answers to typical research questions.

When you listen carefully, instead of just listening to get responses to the questions you've asked, you're directed more easily to follow-up questions or to clarifications. Listening is a major part of your participation in the interview, and showing that you are listening both through intelligent follow-up questions and through body language can be very effective in helping an interviewee respond to your questions.

In addition, watch what's going on, especially if you're gathering information about a process or a procedure. You may need to view the activity several times before you're sure you "get it" and can write about it accurately.

Recording Information

Once the interview gets started, you take notes. But note taking can occur in several ways. If you're taking notes the traditional way with paper and pen or pencil, you want to make sure you can quickly capture the essence of the expert's information. Your notes must be complete enough to make sense to you after the interview. While taking notes, you should maintain eye contact with the interviewee. At all times you should create a friendly, relaxed atmosphere, through conversation, interest in the subject, and eye contact.

You might want to take notes with a laptop computer or a computerized notebook. Again, you should make sure that you take all the notes you need, but you want to be able to participate in the conversation and maintain eye contact. No one feels comfortable during an interview when the person asking questions is so busy recording answers that he or she fails to look up.

Tape recorders or videotape recorders are other options for taking notes. However, sometimes tape recorders make an interviewee uncomfortable, particularly when you and the interviewee work for the same company. Using a tape recorder in this situation may seem like an unnecessarily formal method of gathering information or may violate company rules for confidentiality. But when you're in a situation where you may not be able to talk with the interviewee again to check information, or where the process is better seen than described, you might prefer to use the taping equipment.

Because the interviewee may not be available for additional interviews, or you may not be able to visit the site again, ask all your questions during the interview and check the information for accuracy before the end of the interview.

Setting Up Follow-up Activities

As the interview ends, make any plans to follow up the interview. If, for example, approval of quotations is necessary before you can publish them, before you leave the site, you must establish the process of getting that approval. If another site visit is necessary and can be arranged, set up an appointment before the current interview ends. And always end the interview by thanking the interviewee for taking the time to answer your questions.

Maintaining Confidentiality

Occasionally, researchers conduct formal interviews with people who work with classified information or need to maintain confidentiality about the kind of work they do, the information produced by their company, and the way their answers to interview questions are used. In these cases, you may need to receive a security clearance or sign a confidentiality agreement that requires you to have information approved before it is published or distributed. Interviewees must be assured that the information they provide will in no way jeopardize their jobs or their company's work. Because technical communicators are often working with government agencies, the nature of information gathered during an interview may need to be reviewed to make sure company policies and security measures are not being violated.

Step 4. Follow Up

After the interview, thank the interviewee. Again, a moment of friendly chatting can end the interview in a pleasing way. If you need to send a more formal thank you letter, send it about a day after the interview. A formal letter of thanks to the interviewee is especially important if you conducted a formal interview with someone outside your company. The letter does not have to be long, but it should sincerely express your thanks and note any follow-up appointments or procedures. If you've conducted an in-house interview, or if you're very familiar with the expert, you might simply call or send an e-mail message to express your thanks.

As soon as possible, review your notes or replay the tape and add any comments or descriptions to flesh out the information. You should add explanatory notes so that you will be able to remember what the notes represent. Sometimes you might want to key in notes as an outline on the computer or create a file of notes for future reference. Any of these activities serves to reinforce the information you received during the interview.

Sometimes your follow-up activities include providing a copy of the final document, if your research was designed to gather information to create a specific product. For example, if you're a freelancer writing an article, the interviewee may want a copy of the article when it's published. Or you might need to get permission or final approval for the final form of the information, so you need to complete the appropriate tasks to gain that approval. You might need to have a form signed, or a sign-off line added to the final document. Whatever the requirements might be, complete them in a timely and appropriate fashion.

The follow-up is as important as any other part of the interview process. Perhaps it is more important in some ways, because if you fail to follow up as you said you would, then you may develop a poor reputation as an interviewer. If you follow the niceties, as well as the practicalities, of following up on the interview, you're highly likely to keep the contacts you've made and develop a good reputation as an interviewer, one that experts will want to talk with again.

Conference Calls and Special Situations

Occasionally, the perfect people to interview live at distances that don't lend themselves practically to your time or budget. Perhaps one interviewee works in New York and another in California, and you really should talk with both of them in a collaborative discussion. Under conditions like these, conference calls are useful. A conference call allows more than two people to talk together as if they were in the same room discussing an issue. Conference calls can be a convenient means of talking with several people at once, but as the interviewer you'll probably be responsible for setting up the call and making sure everyone benefits from the call. That requires a great deal of planning and cooperation among the interviewees.

You initially call one person at a time and ask for a convenient time for a conference call. Once a mutually agreeable time has been established, taking into account different time zones and possibly telephone rates at different times of the workday, you call all the interviewees and tell them the time the conference call will take place and the information they should have ready.

During the conference call, you act as a moderator who provides questions, restates information to make sure you understand what is being said, and gives equal time to the interviewees. Because picture phones are not yet in common use, your telephone voice must indicate the warmth and professionalism of a handshake. Of course, listening skills are crucial during a conference call, and you must not only keep the interview on track but comment on what the interviewees are saying. The calls should be efficiently run, too, so that you gather all the important information without running up a huge telephone bill, even if your company is paying for the call at corporate rates.

Taking notes during a conference call may pose a difficulty, but recording conversations without the interviewees' knowledge is illegal, so note taking might be your best option in most cases. If the interviewees agree to be recorded during the call, you can plan ahead for a recording device to be set up in the phone itself so that a clear recording is made. However, many times the interviewees will not want to be recorded, and note taking must suffice.

Conference calls are not the most popular form of interviews, but they are becoming more important in technical communication, as professionals work in all parts of the world. No matter how frequently you use conference calls, though, some people will not like them and will ask you to set up another type of interview. Although the telephone is convenient, some people just don't like the impersonal nature of a conference call and prefer to talk with you in person. Remember, then, as you plan conference calls, that a great part of your job will be in putting your interviewees at ease, even if they can't see you or each other. Instead of acting purely as a researcher in this case, you may need to play the role of friend or facilitator.

Teleconferencing is another option for interviews among several people. Again, you follow the same steps for setting up the interview, and as with a conference call, you facilitate the meeting. Teleconferencing has the benefit of providing visual as well as audio contact, but because it is more expensive and less often used for day-to-day interviewing, you'll have to use this resource carefully if you need it to gather information.

Summary

Good informational interviews require you to use your people skills. As a technical communicator who just happens to be a researcher, you must always remember the communicator part of our profession. By being genuinely interested in the person you're interviewing, you create a warmth that can't be planned but yet is a crucial part of putting the interviewee at ease, developing a rapport so you can easily gather information, and enjoying this part of your job.

But the technical part of our profession also means that you're a methodical researcher. By following the four steps for conducting an informational interview—whether you're conducting one interview or several, on the same schedule or on overlapping schedules, for one project or many, for a job search or as part of your regular job—you know that you'll get the information you need.

Research Activities

1. List three people you need to interview for a research project you're currently working on. Who are they? What are their credentials? How can you get an interview with them? Why are they the best people to interview?
2. Arrange an interview with at least one of the people you listed. Follow the four-step plan for conducting the interview.
3. After the interview, evaluate your performance as an interviewer. What will you do differently next time? Which questions worked well? Which questions should you add next time? How can you

prepare yourself for the next interview? How can you put your interviewee at ease? Did you get all the information you needed? Did you establish rapport with your interviewee? What kind of feedback did you get from your interviewee?

4. Further practice your interview skills by setting up two-person practice sessions in which you are the interviewer and then the interviewee. You might also want to set up a videotaping session and then evaluate your performance during the interview.

For Further Reading

McDowell, Earl E. 1991. *Interviewing Practices for Technical Writers.* Amityville, NY: Baywood.

Slavens, Thomas P. 1985. *Reference Interviews, Questions, and Materials.* 2nd ed. Metuchen, NJ: Scarecrow Press.

CHAPTER NINE

Letters of Inquiry

Even in this electronic age, when sending a letter seems old worldly, formal, and slow, letters of inquiry are still an important research method to gather information. Some people, companies, or institutions require a formal request when researchers are seeking information from them. At other times, a letter of inquiry is the most convenient, or the only, form of communication for the person from whom we request information. Letters of inquiry are used in two specific situations:

- To request information or materials that have been advertised in some way.
- To send an unsolicited request for information or materials.

Sending a letter of inquiry through the regular mails is still considered a time-honored, appropriate manner of soliciting information. In fact, at least perceptually, traditional mail may be considered the most formal way of sending such a letter, and many people still prefer a formal request. Sending letters that request the aid of others (and that's exactly what a letter of inquiry does) through an electronic form such as fax or e-mail may create the perception that the speed of a response is more significant than the completeness of a response. However, people who regularly send and receive fax and e-mail may not have that impression, because they complete much of their business electronically.

Using electronic media to send letters of inquiry can provide you with a number of advantages over traditional mail. Faxed letters of inquiry allow you to

- Keep the original written request for your files, either online or in hard copy.
- Send requests at any time to any location with an operating fax machine.
- Send requests directly from your computer screen to the fax machine, and resulting in a printed copy or an online document.
- Send a form letter to several different people or companies in different locations at precisely the same time.
- Learn immediately if the faxed copy was received (even though you can't guarantee that the person to whom you wrote will read it immediately).
- Anticipate a returned response in the same day.

E-mail is another fast way to send a request for information. Technical communicators might use e-mail to

- Receive a quick answer to a question.
- Obtain permission to download information or permission to use information found on a bulletin board or an e-mail message.
- Order information in paper form.
- Subscribe to online mailing lists, where they can request more information.

Of course, some people who have the information you need may not have access to electronic communication or may prefer the time-honored tradition of making a formal written request. E-mail messages or a faxed letter may seem too hurried, as if you hadn't planned carefully when you needed the information or as if your request was a last-minute consideration. Although you may prefer electronic means of communication, others may not. You complete the same kinds of audience analysis about the recipients of your letters of inquiry as you do when you write information for the users of your carefully researched and designed projects.

Guidelines for Writing Letters of Inquiry

Regardless the medium you may use for sending a letter of inquiry, certain rules for format, content, and style apply. These rules may vary, depending upon the medium, but you still conduct an audience analysis to learn which format, type of content, and style your audience expects. By meeting your readers' expectations, you're more likely to receive a response, especially a timely, thoughtful one.

Format

Letters may be formatted in a number of ways, for example, in a block or a modified block style, or a variation on one of these basic formats. Variations on the modified block style include indenting paragraphs approximately a half inch and leaving a blank line between paragraphs and indenting paragraphs but leaving no line blank between paragraphs. You may decide which format to use, or your company may require a specific format. Figures 9.1 and 9.2 illustrate the basic components of a letter in block and modified block formats.

Content and Organization

When you write a letter of inquiry, you're asking someone to take the time both to read the letter and to respond to it, at a minimum. Sometimes that response can become more involved than just writing answers to questions. If the recipient of your letter has to do some research before he or she can give you answers, copy materials to send to you, or just package and send materials, the response is taking quite a bit of that person's time. As a requester of information, therefore, you should make it easy for the person to respond to your request and indicate the advantages of responding to the request.

You can ease the response process by clearly stating in a logical order who you are, why you are writing to that person in particular to solicit information, and exactly what information you need. When it's appropriate, you further ease the response by providing such courtesies as self-addressed, stamped envelopes.

The advantages to the responder for replying to your request may include a courteous note in the product that results from your research (such as, an acknowledgment in a research article), a copy of the results of your research, or a simple thank you, with an offer of your assistance should the person need it for his or her research.

Your Street Address (when you're not using letterhead)
City, ST Zip Code

Today's Date

Inside Address (to the person who'll receive the letter):
Recipient's Name,
Recipient's Title
Company
Street Address
City, ST Zip Code

Salutation (e.g., Dear ____: or Dear ____,)

1st ¶ (flush with the left margin)

2nd ¶ (flush with the left margin)

Any other ¶ (flush with the left margin)

Closing (e.g., Sincerely,)

Your Printed Name
Your Title (used as necessary)

Enclosures (used as necessary, e.g., encl:)

Copy lines (used as necessary, e.g., cc:)

1234 Main Street
Hometown, OH 12345

January 1, 1995

Jane Smith
9876 Main Street
Hometown, OH 12345

Dear Ms. Smith,

Thank you for inviting me to speak with your colleagues during the next luncheon meeting. As you requested, I'll bring my slides from my recent visits to research laboratories throughout the Midwest.

I'll arrive at 11:30 so that I can set up the equipment before the luncheon begins. All I need is a screen. I'll bring the rest of the equipment.

I'm looking forward to seeing you again. If you would like me to bring additional materials for the presentation, please call me at 987-6543.

Sincerely,

Sam Johnson

Figure 9.1 A block format for a letter.

Your Street Address (when you're not
City, ST Zip Code using letterhead)

Today's Date

Inside Address (to the person who'll receive the letter):
Recipient's Name,
Recipient's Title
Company
Street Address
City, ST Zip Code

Salutation (e.g., Dear ____: or Dear ____,)

1st ¶ (flush with the left margin)

2nd ¶ (flush with the left margin)

Any other ¶ (flush with the left margin)

Closing (e.g., Sincerely,)

Your Printed Name
Your Title (used as necessary)

Enclosures (used as necessary, e.g., encl:)

Copy lines (used as necessary, e.g., cc:)

1234 Main Street
Hometown, OH 12345

January 1, 1995

Jane Smith
9876 Main Street
Hometown, OH 12345

Dear Ms. Smith,

Thank you for inviting me to speak with your colleagues during the next luncheon meeting. As you requested, I'll bring my slides from my recent visits to research laboratories throughout the Midwest.

I'll arrive at 11:30 so that I can set up the equipment before the luncheon begins. All I need is a screen. I'll bring the rest of the equipment.

I'm looking forward to seeing you again. If you would like me to bring additional materials for the presentation, please call me at 987-6543.

Sincerely,

Sam Johnson

Figure 9.2 A modified block format for a letter.

Style

Your letter's style should be appropriately formal to reflect the importance of the subject matter and your relationship with the recipient. If you don't know the person to whom you're writing, it is always best to address the person by title and last name (Dr. Jones, for example).

Formality of style may be reflected through the tone you indicate by your choice of words and the precision and correctness of your grammar. Although contractions are used more frequently, even in the most formal kinds of documents, you might avoid using them if they could suggest a degree of informality you don't intend.

As both a courtesy to your responder and a stylistic approach to help ensure that all information you request is provided, you might itemize, perhaps even number, the different kinds of information you are seeking. If you're asking a series of questions, you can number those to ensure that the responder considers each one separately and can easily see all the questions you have (see Figure 9.3).

A Request for Advertised Materials

As a researcher, you may be responding to an advertisement or some other solicitation that asks you to send a request for further information or materials. For example, you watched *60 Minutes*, and during the program, a subject you're researching was discussed for about 10 minutes. You hadn't heard the evidence presented in that segment and need a transcript of the broadcast. (Eventually, you may also need to write for copyright permission to quote the source if you use it in your project.) At the end of the program you saw a notice requesting that people interested in a transcript write to the address shown on the screen. So, you need to write a letter of inquiry to request a transcript of the broadcast.

These letters are easy to write. You simply write to the people named and request the information, something like "Please send me the transcript of *60 Minutes* from [date]. Thank you very much." A good format to follow is this: "Please send me the (whatever it was they said to write about) noted in (whatever source you saw). Thank you very much." The letter in Figure 9.4 shows a slightly expanded version of this basic request. It provides enough information for you and for the broadcasting company to keep your request on file, but it's short and direct and requires little writing time.

English Department
Bowling Green State University
Bowling Green, OH 43403

Albert Smith, Ph.D.
Another University
Anytown, OH 12345

February 27, 1995

Dear Dr. Smith:

I read your e-mail announcement on the bulletin board that you taught an experimental course in markup languages during the Fall semester 1994. I'm currently teaching an experimental course in markup languages here at BGSU. Although I encountered no problems in getting permission to offer the course, I may need to justify it in writing if I decide to teach it again. I'm also frantically collecting materials for the course because it is being offered at the request of some students and was not approved until much too late for me to complete extensive research.

Would you assist me please by sending me any of the following kinds of information you have:

1. course syllabus
2. course reading list
3. list of electronic newsgroups or listservs you find helpful
4. a course proposal, if you were required to write one to get your course approved

Would you also tell me, please,

1. How many students enrolled in the course?
2. Were they first year? sophomores? juniors? seniors? graduate students?
3. Do you plan to offer the course again? Regularly?

I realize that I'm asking for a lot of your time to respond to my requests, and I thank you very much for that time. I will send you a copy of my syllabus, and I'll properly attribute any of your materials that I use in my class or for any other purpose. If you have any questions about my requests or if the material you have can most easily be sent electronically, please feel free to e-mail the information to me at bcoggin@xxxx.xxxx.xxx.

Sincerely,

William O. Coggin
Professor—Scientific and Technical Communication

Figure 9.3 Enumeration within a letter of inquiry.

514 May Lane
Wooster, OH

September 11, 1994

60 Minutes
CBS
Rockefeller Center
New York, NY

Dear Director, Transcript Services,

Please send me the written transcript of *60 Minutes* that was first aired on September 11, 1994.

I have enclosed a check for $20.00 for the transcript, shipping, and handling.

Thank you.

Sincerely,

Bartley A. Porter

Figure 9.4 Letter of inquiry for a transcript.

An Unsolicited Request for Information

The second situation is more difficult to address; unsolicited requests ask someone to take time in his or her day to do you a favor. Therefore, this request must be handled much more carefully. In this situation, you explain who you are, what information you need, why you're writing to that particular person or company, and the benefits of him or her supplying the information. At times there may be no direct benefits for the person or company; a simple "thank you" has to suffice. If you're writing an article for publication or a report, you can tell the person you will name him or her in the article or report, or you can say you will send a copy of the results of your research. In any case, send a stamped, self-addressed envelope for ease in returning the information you request.

Letters of inquiry are especially useful when the person who will receive the letter has materials that can be sent. Copies of corporate reports, in-house publications, reprints of articles, back issues of periodicals, or generally unavailable information may be impossible to find any other way. The person who can send you documents may not be a good person to interview, because the materials are equally important as personal knowledge or because the person cannot be interviewed at a mutually convenient time or place.

The letter should be direct and clearly request the specific materials you need or ask the questions for which you need an answer. Include a deadline for response but be sure to make the deadline reasonable, especially if you're requesting information or materials that may take some time to gather.

The sample letter (Figure 9.5) requesting a copy of an article currently out of print gives you one approach to conducting research through the mail.

October 1, 1994

Michael W. Wiederman, Ph.D.
Department of Psychology
Kansas State University
Wichita, KS

Dear Dr. Wiederman,

I read your article, "Demographic and Sexual Characteristics of Nonresponders to Sexual Experience Items in a National Survey," in the February 1993 issue of *The Journal of Sex Research*. If you have additional information about the research leading to this article, such as preliminary reports or the complete survey you used, will you send one copy of the information? I will send you a check if you let me know the cost of the forms, copies, or surveys.

My students and I are planning a national survey in which we hope to analyze patterns of nonresponse depending upon the selection of interviewers. We hope to learn more about designing surveys for multicultural communities. We would be happy to send you a copy of our results at the completion of our research.

I've enclosed a stamped, self-addressed envelope for the reprints and additional information you might send. Please call me at 419-424-4766 to let me know of the cost of the materials, and I'll send you a check immediately.

Thank you very much.

Sincerely,

Lynnette R. Porter, Ph.D.

Figure 9.5 Letter of inquiry for additional information.

An Unsolicited Letter to Ask Questions

Sometimes you need answers to questions instead of original or copies of documents. You can vary the format of the letter only slightly when you need to ask questions of a professional who has specialized knowledge. Your letter will be formal, and you should enumerate or highlight your questions so that they're easy to see within the letter. Your questions should be clearly focused and specific, and you should limit the number of questions. If you briefly explain the reason why you need the information, ask three or four specific questions, provide a fast, cost-effective way to send a response, and thank the expert for his or her help, you're likely to get a response.

Faxing a Letter of Inquiry

Faxed letters retain some qualities of mailed letters. They can look like traditional letters and can be more formal. If the recipient has a fax machine with high-quality paper, the faxed copy can be saved for records, just like a regularly printed letter. The style and format are the same as those used in mailed letters, because both faxed and mailed copies are designed to be letters.

If you just need a response to a question or a confirmation that original documents will be sent, a fax can help you greatly with your research. You gather the information quickly and plan if you need to locate additional information from other sources. If you need a draft of a document, but not the actual documents, faxed samples can provide you with the models or content that you need for your project. Even if the actual document will be sent in the next few days, you get a head start on using the materials by working with the faxed copies. If you're faxing letters of inquiry, you should follow some basic guidelines as shown in Figure 9.6.

E-mail and Letters

When you write an e-mail message, it is usually short and preferably friendly. People who write e-mail messages often assume that the network creates one big, friendly family, and the formalities of addressing their audience may be forgotten. Instead of writing "Dear Dr. Allgeier,"

Cliff Logan
The University of Findlay
1000 North Main Street
Findlay, OH 45840

January 9, 1995

Carl Meyer,
Product Manager
Trucking Parts Co.
9876 Trucking Drive
Smithton, OH

Dear Mr. Meyer:

I am writing to you because I am researching the trucking industry and electronic communications for an in-class report. I have learned that your company has members representing a number of parts suppliers and manufacturers. Will you answer a few questions about the use of electronic communications in your company?

1. If you currently use a fax machine to communicate with suppliers and manufacturers, how often do you send a fax?

2. If you currently use a fax machine to receive information from suppliers and manufacturers, how often do you receive a fax?

3. How has a fax machine changed the way you do business?

4. If you currently use e-mail to communicate with suppliers and manufacturers, how often do you send messages?

5. How has e-mail changed the way you do business?

My report is due on February 15. I will be happy to send you a copy of my report when it is completed.

Thank you.

Sincerely,

Cliff Logan

Figure 9.6 A letter of inquiry to solicit answers to questions.

for example, many researchers are tempted to write "Hi, Betsy," or—worse—perhaps just "Betsy—." If you're requesting information via e-mail, you need to be more formal than you may usually be, especially if you're communicating with someone you don't know personally.

Another temptation in requesting information through e-mail is to make the message either long and rambling to explain what you need and why you need it or so short that the style is choppy. A good request for information includes enough information so that the recipient can understand what you need and decide how to respond. But it also should be limited to one screen so that your reader doesn't have to move from screen to screen to understand the whole message.

It should go without writing—but we'll write it anyway—that e-mail messages should be planned, just as letters are planned. Haphazard messages, sometimes with grammatical or spelling errors, are not a good way to convince people that you're a reputable researcher or an effective technical communicator. A request that's direct but easy to read and understand because it conforms to the standards for written business communication is more likely to be understood and complied with. Good netiquette for e-mailed requests for information should follow the guidelines found in Figure 9.7. A brief message, shown in Figure 9.8, provides you with one example you can practice using the next time you request information via e-mail.

- Be polite.
- Use a formal form of address for someone you don't know or don't know well.
- State exactly what you need or what you're asking.
- Divide the information so that it has space between chunks but ideally takes up less than one screen for the entire message.
- Use formal grammar in your writing. (Or conversely, avoid using acronyms, abbreviations, and graphics, such as :-> for a smiling face.)
- Proofread the message for grammatical correctness.
- Write your return e-mail address, in case the From message line is difficult to locate or read.
- Avoid cutesy signatures.

Figure 9.7 Guidelines for writing an e-mail letter of inquiry.

To: Terri Hudson, J. Wiley & Sons
From: Lynnette Porter
RE: Request for Style Manual

Dear Ms. Hudson,

Please send me a copy of the style manual your company requires writers and editors to use, or if you use a standard style manual, please let me know what it is, and I'll buy a copy if I don't already have it.

Thank you.

Lynnette Porter
lporter@opie.bgsu.edu

Figure 9.8 An e-mail letter of inquiry.

If you subscribe to a mailing list for technical communicators or SMEs in a particular technical or scientific area, you have access to numerous potential people who can answer your questions or provide materials. If you are making a request to all the subscribers in a mailing list, you may need to provide some background information about you and your research. But you still want to keep the message short and formal—but friendly. These open requests for information often generate quite a number of responses, and they also may start a new conversational thread on the network as people discuss your questions or request. You might vary the request for style manuals, for example, if you're writing an open request to hundreds of subscribers at once, as shown in Figure 9.9.

If you send messages to a list, be sure to write a concise, precise header to indicate what the message is about. Because subscribers often receive hundreds of messages a day, you want yours to stand out and be read, not sent to the electronic trash. The default format for most e-mail messages is to look like a memorandum. It includes To, From, Subject, and CC lines, for example. However, you can still write a letter, even if you're required to provide memorandum-style information as part of the mailing address format for your e-mail system.

After you've received responses, a general thank you sent to subscribers is a good way of noting that you received the responses and are grateful for the assistance. You might even be asked to share your information with the others who subscribe to the list.

To: Subscribers of tech wr-list (You'll probably include the address of the list, which will automatically be included as part of the message.)
From: Lynnette Porter
RE: Request for Style Manuals

I'm researching the types of style manuals used by technical writers and editors in different publishing companies and in-house technical publication departments. I'll use the style manuals and responses I receive in my technical communication classes' unit on style.

What style manual(s) do you use? If you use an in-house document, please write to me directly and let me know if and how I can get a copy.

Thank you.

Lynnette Porter
lporter@opie.bgsu.edu

Figure 9.9 E-mail requests to members of a mailing list or special interest list.

Summary

Writing letters or messages of inquiry requires some finesse. Letters requesting previously advertised or commonly available information are easy to write, and you may develop a form letter that you can routinely use for these requests. In contrast, unsolicited letters requesting a favor from someone who doesn't know you are trickier to write and should be carefully composed and revised. In these letters, the tone, word choice, and appearance of text either can ensure that you will get the information you need on time or can doom your letter to a forgotten stack of papers. If you use e-mail, you need to keep a sense of formality and write a personalized but direct message to the person who can provide you the information you need. The letters or messages are short, but what you say in them can affect your entire research process and the success of your project.

Research Activities

1. Write to a professional society, community agency, federal agency, or other group that publishes reports, booklets, or pamphlets about your research topic and request a copy of one of their publications.

2. Write an unsolicited letter of inquiry to a professional who works with the kind of information you're researching. (To find the name and address of such a person, look in directories listing members of professional societies, managers of large corporations, and speakers in your school's speakers bureau. These types of directories are usually kept in the reference section of the library.)
3. Write an e-mail message of inquiry to a professional who works with the kind of information you're researching. (To find the name and address of such a person, look in directories listing members of professional societies, managers of large corporations, and speakers in your school's speakers bureau. You might also subscribe to a list server with a technical communication mailing list. These lists are used by professionals who converse via e-mail.)

For Further Reading

Dragga, Sam. 1991. "Classifications of Correspondence: Complexity versus Simplicity." *Technical Writing Teacher* 18(1) (Winter): 1-13.

Hall, Dean G., and Bonnie A. Nelson. 1987. "Initiating Students into Professionalism: Teaching the Letter of Inquiry." *Technical Writing Teacher* 14(1) (Winter): 86-9.

___. 1988. "The Letter of Inquiry: A Neglected Tool for Technical Classes." *Engineering Education* 78(7) (April): 695-99.

CHAPTER TEN

Questionnaires and Surveys

Sometimes you can't interview all the people you'd like to, perhaps because there are too many of them, they live in widely separated geographic locations, you don't have enough time, or you want anonymous responses to your questions. Sometimes you need a wider "mix" of people for a survey, instead of focusing on people you know are experts about a subject. Sometimes you may want to target people with certain attributes—they live in a specific geographic region, hold a certain kind of job, are in a specific age group, for example. Therefore, when you want to

- target a certain group who you can't interview,
- obtain a varied sampling of responses from people coming from different backgrounds and experiences,
- solicit many responses, not just a few,
- compare responses from several people,

consider using a questionnaire or a survey to get the information you need.

Questionnaires and surveys share some general characteristics of design and application; in fact, questionnaires are often referred to as surveys, and vice versa. However, there are some differences between the two.

A *survey* is less formal than a questionnaire; it may include only a few questions and is more likely to be "distributed" over the telephone, in person, or via e-mail. A *questionnaire* is more formal; it usually includes more questions and requires more time to complete. Generally, it requires respondents, the people completing the questionnaire, to fill out answers on a paper form or to have someone administering the questionnaire complete the form for the respondents.

Questionnaires can also vary in their purpose. Some business or technical questionnaires may be distributed simply to gather as much information as possible; they're used as learning tools, to see what people think or do and how they respond. These questionnaires should be well designed, but they may not have been empirically tested prior to their distribution. They were not designed as scientific research tools.

Other questionnaires *are* empirical research tools. They've been designed to test a hypothesis; they may be or may become valid, reliable test instruments that become standardized within a discipline. Other researchers may use them as part of their research design, because the questionnaires have proven reliability in certain testing situations. If you're designing a questionnaire as part of your hypothesis testing, and using a questionnaire is your experimental method, you'll have to follow the principles of designing an experiment when you write the questionnaire and determine the sampling method and sample size.

In this chapter, you read about general principles in writing, administering, and collecting data from a questionnaire. If you're designing a questionnaire to be used as an experiment, you'll also want to read Chapter 11, Empirical Experimentation, as part of your preparation. You'll also develop a vocabulary of the types of questions used in questionnaires.

Considerations for Using Questionnaires

Questionnaires take time to plan, distribute, collect, and evaluate, so you should be sure you want and need to use a questionnaire as you plan your research strategy. A poorly written questionnaire just wastes your and your respondents' time and doesn't allow you to collect the information you need. However, when a questionnaire is well designed, it provides you with valuable information you can't get from any other source.

Some positive features of using a questionnaire include the following:

- Respondents have time to think about the questions and their responses.

- Respondents are more accessible by mail than they are for a personal interview. They may also be more accessible in controlled locations, such as a classroom, a factory lunchroom, or a meeting.
- Questionnaires may eliminate the problem of geographic limitations.
- Questionnaires may be completed anonymously, and respondents may be more likely to answer questions if their identity is not known.

Of course, like any other source of information, questionnaires pose the potential for problems in your data gathering. The following are a few potential problems:

- It may be difficult to follow up on answers to questions if the answers are unreadable or if the responses were made anonymously.
- Respondents who have strong feelings about the topic are more likely to reply than those who don't.
- If you're using a mailed questionnaire, the process of mailing questionnaires and waiting for responses can be very time consuming.
- If questionnaires are too long, few replies can be expected.
- If questionnaires are confusing to respondents, either because the instructions aren't precise or because the questions are worded ambiguously, few replies can be expected.

You can minimize some of these disadvantages by carefully designing the questionnaire.

Types of Questionnaires

Questionnaires are useful instruments for collecting various, specific kinds of data and interpreting your findings. Like all other data-collection instruments, however, they have to be designed according to particular specifications and with specific aims in mind. For example, you can design questionnaires that result primarily in data which you can count. That is, you ask questions like "How many?" or "How often?" Or you seek interpretations or impressions. For example, you ask questions like "Why?" Both approaches can provide varieties of data that can be viewed as responses to specific questions but can also be examined in terms of their relationship to other data. Because responses to questionnaires can often be evaluated statistically, you may consult with a statistician to help you accurately interpret the data.

In your work as a technical communicator, you may work with both questionnaires you develop and questionnaires that others develop. As you develop your questionnaires, you determine what kind of information you need to gather and how you can best get that information. If your research is based on someone else's or is designed to replicate someone else's research, you may use a questionnaire that another researcher developed and tested previously. You may find data supporting the questionnaire's reliability and validity and its effectiveness as a research tool. If you use someone else's questionnaire, you should check to ensure the following:

- The instrument has acceptable reliability and validity for your research.
- The instrument can be applied to the experiment that you've developed.
- If you vary the questions, use a shortened form of the questionnaire, or otherwise modify the questionnaire, the revised version is tested and deemed acceptable for your experiment.
- You can get permission to use the questionnaire.

Variables to Consider in Designing a Questionnaire

When you design a questionnaire, as much as you possibly can, anticipate the kinds of conditions that might affect the quality and quantity of data you get. Then, you either try to control those conditions through design and administration of the questionnaire, or you consider them when you interpret the data. The greater control you have over planning the questionnaire, the more likely you are to get valuable results and a large response rate. Realize, though, that no matter how carefully you plan the questionnaire and its use, you can't control all the conditions that affect respondents as they complete the questionnaire. For example, you can't control how people will feel when they open a mailed questionnaire. Maybe someone who had a bad day won't bother completing the questionnaire. Other people may be in a happy, optimistic mood, and their responses reflect that current mood. Some people may want their responses to reflect the way they'd like to be or what they think others expect or prefer; they may indicate responses that don't reflect what they actually do or think. If they know that society frowns on a certain behavior or that an idea is not socially acceptable, they may not want to admit their actual behavior or attitude.

Participants Who Complete the Questionnaire

One of the most important considerations in using a questionnaire is your choice of participants. The people you choose to survey should reflect your research objectives. You want to ensure that you have a large enough sample, or group, because many people do not complete questionnaires or they fail to provide usable answers. You need to determine how representative your sample will be or how randomly you want to choose participants. Also, depending upon the type of research you're conducting, where your respondents are located, and for whom you're doing the research, you may need to get approval for using people in your research. As with other types of experiments, you may need approval from a human subjects review board or another approving agency or group to ensure that your treatment of participants is safe, confidential, and ethical. (A table of sampling methods is included in Chapter 11, Empirical Experimentation.)

Decisions to Make Before You Write the Questionnaire

Technical communicators seldom rely only on questionnaires to gather information, because other methods of information gathering also help them to support or develop a hypothesis or provide data for further research. However, that doesn't minimize the importance of questionnaires; they're useful primary sources of information that can enhance findings from other sources. Answering the following series of questions will help you analyze your research strategy to determine if, and how, the questionnaire fits into your overall plan for gathering information.

1. What are primary and secondary methods of data gathering, and into which category does the questionnaire fall?
2. How will possible respondents be approached (e.g., purpose of the research, confidentiality, anonymity)?

3. How will questions be built into sequences (e.g., the order of the questions, other considerations about the framework of the questionnaire)?
4. What is the order of questions within question sequences, with consideration to the variables you've already analyzed?
5. Will questions be precoded or free response?

After you've answered these questions, you should have a good plan for the way the questionnaire will be structured and also an idea about the number and types of questions you have to ask.

Six Basic Types of Questions

Once you've decided if you want open-ended or closed questions, or in what proportion you'll include some of each type, you must think about specific questions. *Closed questions,* for example, come in many forms, although each form limits the possible number of responses. *Open-ended questions* allow respondents to complete the item in their own words. The following six types of questions provide you with a nice variety of selections to meet your needs for responses and the audience's needs for understanding the questions and quickly completing the questionnaire.

- Dichotomous
- Multiple Choice
- Ranking
- Rating
- Fill-in-the-blank
- Essay

Dichotomous Questions

Because *dichotomy* means "division into two separate parts," these questions provide the respondent a choice between two answers (such as Yes or No). They are a form of closed questions. For both these reasons, dichotomous questions are usually good warm-up questions for respondents. Most people can answer these types of questions quickly and without much thought. Although the amount of information you receive from dichotomous questions is limited by the nature of the question, you can gather a number of responses fairly quickly.

Multiple Choice Questions

Multiple choice questions provide several possible answers to choose from. They are a form of closed questions. Multiple choice questions are easier for respondents to answer if you limit the number of choices, but still provide an adequate variety of responses. (Remember: You want to make the questions as unbiased as possible. Don't skew the question by including a narrow range of responses.) For example, you may remember a test you took in school that included choices labeled A, B, C, and D. But the responses went on. They also included E, all of the above; F, none of the above; G, A, and B; H, B, and C; I, A, B, and C only; and so on. By the time you finished reading the list of responses, you were thoroughly confused. Respondents who see more than four or five possible choices in response to a multiple choice question probably will move on to the next question. Usually, you can provide an appropriate range of typical responses within four or five response items.

Ranking

A ranking question lists several possibilities and requires respondents to indicate numerically a hierarchy by preference (such as Who do you prefer among the following instructors to teach this class?) or by frequency (such as In order of frequency, which brand of toothpaste do you use?). Ranking questions are a form of closed questions. As with multiple choice or rating questions, you need to remember that respondents' short-term memory may only be able to handle five to seven items at one time. Although you want a representative list of choices for respondents to rank, you also want to provide them with as short a list as possible.

When you ask respondents to rank a list of items by preference for a product or a behavior or frequency of a behavior or an occurrence, you need to provide them with additional instructions. Some respondents may assume that if they rank an item with a 1, you'll know that 1 indicates their first choice—most frequent behavior or occurrence or most preferred product or behavior. Other respondents, however, may think that the higher the number, the more frequent or more preferred the product. To these respondents, the highest number may be the item they rate most important. Your instructions, then, indicate the way you want respondents to indicate the hierarchy (from 1 to 6, rank the following items in order of your preference for the product. 1 = Most Preferred, 6 = Least Preferred).

Rating

Rating also lets you lists several possibilities, but it requires respondents to select a category of response among a series of possible responses (such as How do you rate our service? Excellent, Good, Average, Poor). These questions are closed questions; you determine the range of answers from which respondents select the appropriate rating. Most ratings include three to five items, but some, such as those on a Likert scale, may include more items. The range of responses is incremental along a continuum, such as Excellent, Good, Fair, Poor; Completely Agree, Agree Somewhat, No Opinion, Disagree Somewhat, Completely Disagree; or Always, Frequently, Sometimes, Seldom, Never. These responses can refer to quality of a product or service or to the frequency of a behavior or an occurrence.

With rating questions, as with multiple choice or ranking questions, you arrange the possible responses so that respondents can see all their options at a glance. Although the number of responses may be limited to a manageable number just by the nature of the continuum, you use the range consistently to ensure that respondents can always find the category that's appropriate to their response. For example, in most questionnaires, you find that the most positive or the most frequent category of response is placed at the top or in the leftmost position of the list of the range of responses. The next most positive or most frequent category is placed below or to the right of it, and so on, as shown in Figure 10.1. Respondents can easily see the range of responses and can more efficiently complete their rating.

Fill-in-the-Blank Questions

Fill-in-the-blank questions guide the respondent to a type of answer, such as a number, place, or date (as in How long have you been a student? _________). They are a form of open-ended questions. These questions are most effective when you indicate in the question the type of response you need. For example, if you ask, "How long have you been a student?" you might provide this place for response: _________ semesters. Then the respondent knows that you're looking for a number of semesters. If you ask, "What percentage of your trash this week do you

Dichotomous

Have you purchased a PowerBook in the past year? Yes No

True or False. A quark is a subatomic particle.

Multiple Choice

When you enter data on Form 1003-A, you first have to _____.

A. enter the client's last name
B. enter the client's Social Security number
C. enter the client's vendor ID number
D. enter the client's address

Ranking

On a scale from 1 to 5, rank your preference for the following food items offered at the lunch counter each day. 1 = most preferred, 5 = least preferred

_____ hamburger
_____ cheeseburger
_____ hot dog
_____ salad
_____ hot vegetable plate

Rating

How do you rate the in-house desktop publishing service on the quality of the materials they've produced for you?

Excellent Good Fair Poor

How often do you purchase software from our company?

Every month

Every 1 to 6 months

Every 7 to 11 months

At least once a year

Every few years

Never

(continues)

Figure 10.1 Examples of questionnaire questions.

Fill-in-the-Blank

Where did you purchase this product? ____________

In the past six months, I have visited the dentist _____ times.

Essay

What are three suggestions for increasing safety within the plant?

1.

2.

3.

When and where should the next meeting be held?

Figure 10.1 Examples of questionnaire questions. (*Continued*)

expect to consist of paper, in any form?" the respondent knows to supply an approximate percentage. Of course, some respondents may be overly precise by providing answers like 3.45 semesters or 72.9183 percent. However, by inviting respondents to determine an appropriate response for them, you receive a wide variety of responses that reflect the respondents' thoughts at that time.

Essay Questions

Essay questions are a form of open-ended questions, but they should guide the respondent to a type of response (as in, "Describe how the workshop could have been better," or "What are three topics you'd like to see in the winter luncheon discussion series?"). Some respondents don't like to answer essay questions because they take more time or thought. They may avoid them because they want to remain anonymous and don't want their ideas, especially if they're critical, to be traced to them. They may find writing difficult and avoid situations in which they have to write. But for all those respondents who don't complete essay questions, other respondents take the time and effort to provide you with useful, insightful comments.

To get more of those insightful comments, you limit the number of essay questions to those questions for which you really need the respondents' own words. Then you direct the question so that respondents have something specific to answer. For example, a question like "How would you improve the sales program?" is much more focused and likely to generate responses than just providing a heading like Comments or Suggestions.

Because essay questions take longer to complete, you may include them at the end of the questionnaire. Some respondents take the space allocations provided under questions very seriously, and if you provide a half page for a response, they may feel that they have to fill the half page. That may be a daunting task. Therefore, make sure you allow enough space for an adequate response, but not so much as to suggest that you're literally requiring an essay.

Selecting Questions

Figure 10.1 provides some examples of different types of questions, from which you can create the appropriate sequences for your audience. When you choose the type(s) of questions you'll use in your questionnaire, remember these guidelines:

- Try to use all questions of the same type at one place in the questionnaire. (Skipping from dichotomous to essay to multiple choice questions as Questions 1, 2, and 3 can frustrate the respondents, and it suggests poor planning.)
- Organize the questions so that the respondents warm up to answering questions. You might give them a few easy questions at first.
- Choose the type(s) of questions that will provide you with the best kind of information. You don't have to use all six types of questions in one questionnaire.

Planning the questions takes some time, and you'll probably find out that your questionnaires are better when you make a rough draft, or several drafts, as you determine the best questioning strategy to use for your target audience. Even after you've carefully planned the questions and the question sequences, you should test the questionnaire with at least a few groups of volunteers or samples of typical respondents before you administer it.

Question Sequences

Because questionnaires consist of questions or question sequences, after determining your research objectives and selecting your group to question, you consider the kinds and order of questions. As you create questions, you must remember your target group, the people who have to answer these questions. Each respondent must be able to understand what you're asking and how you want him or her to record the response to each question. The better you do the initial job of audience analysis, the greater the possibility that your audience will provide you the kinds of data you're seeking.

By now, you've figured out that research methodology involves asking lots of questions. Before you can write questions to ask others, you have to ask yourself another series of questions that will help you determine what you'll write next. The following questions help you analyze the number and type of questions to ask on the questionnaire:

1. Should you start off with easy, nonfactual questions and try to establish rapport? How will these questions affect your target audience?
2. Should you use a funnel approach (that is, start with very broad questions and gradually narrow to very specific points)?
3. Do you want to use filter questions (that is, questions that eliminate certain parts of the audience from certain questions)?
4. Do you want to use open-ended or closed questions? In what sequence should you ask the questions? What kinds of responses do you want to elicit? When you ask closed questions, you provide specific answers that must be marked. When you ask open-ended questions, you don't provide a choice of answers, so you require the respondents to provide answers in their own words.
5. How many questions do you need to ask? How long will it take most respondents to respond to this many questions? How many pages should the questionnaire consist of?

As you answer these questions, always keep your audience—the respondents—in mind. How much time will they have to complete the questionnaire? Are some types of questions easier to answer than others? For example, many people don't like to answer open-ended questions (sometimes called essay questions), because they feel they don't express themselves well or they just don't have the time to answer in great detail.

Therefore, many people leave open-ended questions blank. On the other hand, two-answer questions are the simplest type of closed questions, because they only provide two possible responses (Yes/No, True/False). However, these questions may not help you gather the variety of responses you'd like.

Perhaps another question you should ask yourself at this time is how you will record the number of responses to each question. Will you use a coded answer sheet so that you can have a computer count the number of Yeses or Trues from the group of respondents? Will you have to read each essay answer and manually record responses? As you plan the questions, you should figure out how you're going to reach your audience and how you'll record their responses.

Word Choice in Questions

An accurate sequence and context for the questions, as well as clear, effective wording can make questions easy to understand and as unbiased as possible. You don't want to lead respondents into answering what they know you want to hear or read; you want their own responses. The following suggestions should help you simplify the style of your questions.

- Keep the questions short. Respondents want to finish the questionnaire quickly, and you should get to the point of your question.
- Remember that respondents may not have all the information you need. Let them provide as much information as they know.
- Phrase questions in such a way that the definition of a term is provided for your readers or you can test to see if the respondents really understand the term.
- Remember that people tend to choose numbers that lean toward an average. Also, if you request a numerical response, understand that some respondents will play games with you and will provide a strange number. For example, if you ask for an average number of visits to the dentist each year, some respondent will be tempted to write 15.6.
- Avoid negative questions like "Have you been ill lately?" or "Would you rather not work on weekends?" Phrase questions positively, such as "How have you been feeling lately?" or "When would you rather work?"

- Avoid leading questions like "Do you take breaks only when you're supposed to?" Instead, ask questions like "When do you take breaks?"
- Avoid loaded questions like "Do you consider yourself a trendsetter?" or "Are you above average in intelligence?" These questions are also an example of prestige bias, because many people fancy themselves as trendsetters because of the prestige it implies or believe themselves to be intellectually superior to others because intelligence is a good quality.
- Be tactful when you ask questions that might be embarrassing to some respondents, like "Are you satisfied with your sex life?" You don't have to avoid private matters or subjects that may deal with controversial subjects, but be tactful when you ask questions about these topics.
- Be considerate of your respondents. Provide complete instructions. Use please and thank you in your comments to the respondents. Make the format easy to read and pleasing to the eye. Include a self-addressed, stamped envelope or other easy method for returning the questionnaire by a reasonable date.

You've probably seen or received many questionnaires that were difficult to read, impossible to follow, poorly organized, and formatted poorly. If you're like most people who frequently receive questionnaires through the mail, you might spend about a minute looking at the questionnaire before you decide to complete it or throw it away. If the questionnaire is messy, too long, or difficult to understand, you quickly get tired of figuring out what the researchers want and just won't fill out the questionnaire. Remember those kinds of experiences when you design your questionnaire. These hints should help you use your common sense to create an attractive, informative, well-organized questionnaire.

The Cover Letter

If your questionnaire will be mailed, include a cover letter stating the purpose of the research, the importance or significance of the research, and why it's so important for the respondent to complete the questionnaire. Also include instructions for completing the questionnaire and returning it. Finally, thank the respondent for completing the questionnaire. You may describe how the results will be used or if or how the respondent can find out about the results.

If your questionnaire will be distributed in another way, you still might include some information typically found in a cover letter, but in a different form. If interviewers will ask questions and record responses from respondents, they might read a statement about the questionnaire and then thank respondents who provide information. If the questionnaire is distributed in person, but the respondent fills out the information and returns it to a designated location or person, the explanatory information and appropriate thanks can be written on the questionnaire form itself or provided by the person distributing the questionnaires.

A Sample Questionnaire

The questionnaire in Figure 10.2 illustrates some qualities of a good questionnaire. The format is easy to follow; the questions are easy to understand; and the technical communicator who created the questionnaire has established a strategy of warming up the respondents and increasing the difficulty and specificity of the questions as the respondent answers more questions. A cover letter explains the purpose of the questionnaire, the method of returning the questionnaire, and the way to complete questions. The target audience for this questionnaire is students on a college campus who have access to the community's library.

A more formal questionnaire, sent to a larger group, is the Society for Technical Communication's salary survey, which is sent each year to a sample of technical communicators who are members of STC (see Figure 10.3).

Questionnaire Distribution

After you've designed and produced a good questionnaire, you need a good method of distributing the questionnaires to your target audience. Often, you'll need to distribute questionnaires to people located in widely separated geographic regions, but occasionally your target audience will be in nearby locations. Some typical ways of distributing questionnaires can give you ideas for the best way to reach your target audience. Questionnaires are usually distributed by mail, but they can also be

Evaluating the Public Library

Please check only one answer for each question.

1. Have you visited the public library in the past year? ___Yes ___No

2. Does the public library provide you with different materials than the college library? ___Yes ___No

3. Does the public library have convenient hours? ___Yes ___No

Please check as many answers as apply.

4. Which of the following resources have you used in the public library?
   - ___ the card catalog
   - ___ the online catalog
   - ___ the online periodical guide
   - ___ the printed reference guides
   - ___ the Reference Desk
   - ___ none of these resources

5. Approximately how much of the time do you find the materials you need from the public library?
   - ___ I don't look for materials at this library.
   - ___ 0%
   - ___ 25%
   - ___ 50%
   - ___ 75%
   - ___ 100%

Please answer the following questions in the space provided after the question.

6. If you haven't found all the materials you need in the public library, which materials have been unavailable?

7. Overall, what is your evaluation of the public library as a resource for your college-related research?

Thank you for completing the questionnaire. Please return it in the accompanying envelope.

Figure 10.2 A short questionnaire.

society for technical communication

Annual Salary Survey

Please take a few minutes to complete this questionnaire and mail it to the STC office as soon as possible using the enclosed envelope. (Please mail it no later than November 30, 1994.)

Part I - Your Employment Status

1. What is your employment status? (Check only one.)

______a. Full-time technical writer/editor (i.e., 35+ hours per week); not consultant/independent contractor
______b. Part-time technical writer/editor
______c. Consultant/independent contractor
______d. Unemployed
______e. Retired
______f. Other (please explain) ________________________________

If your answer to question 1 is "a. Full-time technical writer/editor...," please complete the rest of the survey. If not, please stop and return this form to STC using the enclosed envelope.

Part II - Work History

(For purposes of this survey, the terms "employer" and "company" are used interchangeably.)

2. How many years have you worked in the technical communication profession? ______

3. How did you first start working as a technical writer/editor?

______a. I went straight from school to working as a technical writer/editor.
______b. I re-entered the job market and started working as a technical writer/editor.
______c. I stayed with the same employer, but switched from a non-technical communication position to a technical communication position.
______d. I changed employers and moved from a non-technical communication position to a technical communication position.
______e. Other (please explain) ________________________________

4. How many different corporations (employers) have you worked for while in the technical communication profession?_______

1

Figure 10.3 Society for Technical Communication salary survey.

Part III - Present Employer

5. How many years have you worked for your present employer? ______

6. Is your employer the federal, state, or city government?
______a. Yes
______b. No

7. Is your employer a university or college?
______a. Yes
______b. No

8. Is your employer considered part of the computer industry?
______a. Yes
______b. No

9. Is your employer a contracting company specializing in technical communication?
______a. Yes
______b. No

10. How many employees currently work for your company? (Present site only.)
______a. 1-10 employees
______b. 11-25
______c. 26-100
______d. 101-500
______e. More than 500
If possible, please provide an exact number of employees. ____________

11. How are the needs of technical communication services fulfilled in your company? (Check all that apply.)
______a. In-house technical communication staff
______b. Other in-house staff
______c. External consultants
______d. Other (please explain) ____

Part IV - Current Position

12. What is your job title? ____

2

Figure 10.3 Society for Technical Communication salary survey. (*Continued*)

13. How would you describe your employment level?
______a. Entry-level
______b. Mid-level, non-supervisory
______c. Mid-level, management
______d. Senior-level management, non-supervisory
______e. Senior-level management
______f. Other (please explain) ______________________________

14. Do you write original material or rewrite/edit someone else's material?
______a. Write original material
______b. Rewrite/edit
______c. Some of both
If both, approximately what percentage of your time writing/editing is spent writing original material? _____%
______d. Neither

15. Were specific computer skills (e.g., Ventura, Macintosh, Windows, etc.) required as a condition of your being hired for or moved to your current position?
______a. Yes
______b. No

16. How important were the following skills to your employer in your being hired for or moved to your current position? (circle the appropriate response.)

	Very Important	Somewhat Important	Slightly Important	Not at All Important
Specific computer skills	1	2	3	4
General computer skills	1	2	3	4
Writing skills	1	2	3	4
Editing skills	1	2	3	4
Managerial skills	1	2	3	4

Part V - Salary/Fringe Benefit Information

17. How are you compensated?
______a. Salary
______b. Hourly wage

- **If you are salaried,** what is your current annual base salary? (Do not include the monetary value of fringe benefits.) $ ____________

- **If you are paid by the hour,** what is your current hourly wage? (Do not include the monetary value of fringe benefits.) $ ____________

3

Figure 10.3 (*Continued*)

18. How many hours per week, on the average, do you customarily work? _________

19. What benefits do you receive in addition to salary? (Check all that apply.)
______a. Health insurance
______b. Life insurance
______c. Retirement/pension
______d. Dental insurance
______e. Disability insurance
______f. Professional Society dues
______g. Seminar/conference reimbursements
______h. Internet access
______i. Other (please explain)________________________________

Part VI - Demographic Information

20. What is the highest level of formal education you have completed?
______a. Doctorate
______b. Master's degree
______c. Bachelor's degree
______d. High School diploma

21. What is your sex?
______a. Female
______b. Male

22. What is your age? _______

23. What is your zip code? _______________

Thank you for participating in this survey.
Please return to STC using the enclosed envelope.

Society for Technical Communication
901 N. Stuart Street, Suite 904
Arlington, VA 22203-1854

4

Figure 10.3 Society for Technical Communication salary survey. (*Continued*)

delivered by hand or kept in one location, depending on the researcher's strategy for gathering information, the time frame for gathering data, and the location of the target audience. You can also distribute the information in different ways, through printed questionnaires that respondents mark, printed response sheets that accompany the questionnaire, e-mail responses to questionnaires distributed electronically, and oral responses to questions and the choices of responses that an interviewer reads, either over the phone or in person.

Mailed Questionnaires

Respondents who are located in widely separated geographic areas, who work in different companies, or who are otherwise difficult to reach usually receive and return their questionnaires through the mail. Mass mailings are often sent at a bulk rate, and you can be reasonably certain that respondents will receive the questionnaire. By providing respondents with a free, easy way to return the questionnaire, most often with a stamped, self-addressed envelope, you simplify the information-gathering process.

In-Person Distribution

Sometimes you, perhaps assisted by colleagues or trained volunteers, want to survey a group in one location at one time and can distribute the questionnaires in person. If you conduct research among workers in a factory, for example, you might distribute a questionnaire to each worker entering the building and ask that the respondent return the questionnaire by a specified time (such as before the end of the shift) at a specified place (such as a collection box by the front door).

Market surveys are occasionally done this way; questionnaires are distributed at the door of a fast food restaurant, and the patrons are asked to complete the questionnaire and return it to the researcher before they leave. Distributing questionnaires by hand is a quick way of giving the questionnaires to a target group in one location; it isn't good for mass distribution.

Distributing Questionnaires at One Location

Some businesses have an ongoing use for questionnaires. They therefore keep questionnaires on hand in one location. For example, restaurants that are part of a chain sometimes provide questionnaire cards on the tables so that patrons who have the time and interest in filling out a card may do so. The cards are then kept and reviewed at certain times of the year. Car dealerships also keep questionnaire forms in stock so that new car buyers can evaluate the sales representative's performance and the quality of the dealership's service. Again, periodically these forms are evaluated.

E-mail Questionnaires

With the availability of e-mail, many technical communicators are taking quick, informal surveys of their colleagues' or experts' ideas and opinions. Although these questionnaires may not be as formally prepared, they should still be well thought out before they're distributed. If you're considering surveying a group through e-mail, whether you select individuals and mail them directly or put your questionnaire out to a newsgroup or mailing list, you should follow many of the same conventions you use in a mailed questionnaire: describe the research, carefully design the questions, provide instructions, and indicate when you need responses. You might also indicate that the findings will be made available on the network as soon as you evaluate the responses.

Recording Responses

Respondents who write answers to the questions may mark the questionnaires or record their responses on computer-coded response sheets. If you distribute the questionnaire electronically via e-mail, respondents may send e-mail messages in response. The form of the information may then need to be coded or recorded in another way so that you can compare responses received from different people.

Sometimes a group of respondents may not be able to read and respond to a printed questionnaire. In this situation, you may need to read the questions and responses and record the respondents' answers. You may ask questions at the respondent's home, place of work, your research site (such as a research center), or at another neutral location. If interviewers, including you and any colleagues or volunteers who assist you, are going to ask or read the questions and possible responses, they should be trained so that their body language, facial expressions, vocal inflection, or interaction with the respondents remains as neutral as possible. Consciously or unconsciously, interviewers shouldn't bias the questions or lead respondents into answering questions in a certain way.

Evaluation of the Responses

In the planning stages, you decide on the way you or someone else will evaluate the responses you receive. If you need a valid, reliable questionnaire resulting in data to be evaluated statistically, you'll probably design questions that can easily be coded and perhaps read by a computer that will count the number of responses to each question and break down the responses statistically. If you need individual responses or a variety of answers given in the respondent's words, you'll probably include essay questions that need to be read and evaluated by people. The questionnaire's design and your method of receiving the data should facilitate a quick and efficient method of evaluating responses and interpreting the findings. If you include open-ended questions, you especially need a list of criteria by which you evaluate responses or categorize the comments you receive.

Statistical experts can provide you with information about statistical analysis or validating the responses you receive. (You may want to refer to Chapter 13, Statistics and Measurements, for definitions and brief descriptions of the tests and terms used to evaluate your research.) To receive help in evaluating your questionnaires, you should discuss the types of questions and your method of gathering responses with a professional statistician or data analyst who can tell you the kind of assistance you can receive in evaluating and interpreting questionnaire responses.

Summary

As a quick review, here are some important ideas you should remember about questionnaire design:

- Include only one idea in each question.
- Make questions easy to answer.
- Avoid questions and suggested answers that prejudice the result.

- For each objective you are trying to accomplish, make a new questionnaire.
- Keep questionnaires as short as possible.
- Check the connotations of words used in questions.
- Keep questions as specific as possible.

By breaking down the tasks involved in creating a questionnaire, you can quickly manage the planning, development, distribution, and evaluation of questionnaires and the data they generate.

Research Activities

1. Plan a questionnaire that can be distributed on campus to students and/or faculty and staff. Limit the questionnaire to one page and set a deadline for gathering information at one week.
    - What is the best method of distributing and gathering questionnaires in such a short time?
    - How can you get the potential respondents interested in completing the questionnaire?
    - How can you phrase the questions to get the information you want from such a diverse population?
    - When you're limited to one page, how many questions can you place on that page without cluttering it?
    - Which questions are the most important, when you're limited by space?
    - How will you evaluate the responses?
2. Evaluate the usefulness of a questionnaire in gathering information for a current or anticipated research project.
    - Is a questionnaire a good way of gathering information? Why or why not?
    - If you plan to use a questionnaire, who is your target group?
    - How much time do you have to develop the questionnaire?
    - How much money do you have for producing and distributing the questionnaires?

- What facilities are available to you for producing, distributing, and evaluating the questionnaires?
- If you're going to use a questionnaire, is it the only method you'll use for gathering information?

For Further Reading

Bell, Judith. 1993. *Doing Your Research Project: A Guide for First-time Researchers in Education and Social Science.* 2nd ed. Philadelphia: Open University Press.

Cragg, Paul B. 1991. "Designing and Using Mail Questionnaires." N. Craig Smith and Paul Dainty, eds. *The Management Research Handbook.* New York: Routledge.

Jobber, David. 1991. "Choosing a Survey Method in Management Research." *The Management Research Handbook.* N. Craig Smith and Paul Dainty, eds. New York: Routledge.

Newell, Rosemarie. 1993. "Questionnaires." *Researching Social Life,* pp. 94-115. Nigel Gilbert, ed. Newbury Park, NJ: Sage.

Plumb, Carolyn. 1992. "Survey Research in Technical Communication: Designing and Administering Questionnaires." *Technical Communication* (November): 625-38.

Robson, Colin. 1993. *Real World Research: A Resource for Social Scientists and Practitioner-Researchers.* Cambridge, MA: Blackwell.

Spring, Marietta. 1988. "Writing a Questionnaire Report." *Bulletin of the Association for Business Communication* (September): 18–19.

CHAPTER ELEVEN

Empirical Experimentation

As you read journals or chapters in books, regardless of the subject matter with which they are concerned, you may notice a rather consistent organizational pattern in the articles. In some cases the pattern may be distinctly noted through headings; in others, you may discover the pattern through following the flow of the content. For example, usually there's an introductory section that provides background information about the research. Then, particularly if the article is based on other publications, you may find a section that discusses other literature written about the same or related subjects. This literature review might be part of the introduction or it might be a separate section.

Following this background information, you may find a section devoted entirely to providing a statement that notes the objective or purpose of the article you're reading. This statement, sometimes called the thesis, provides the central point that the remainder of the article is designed to support.

When it's appropriate, for example, if the exact procedures the authors used to support the point they're making are essential for the reader to believe the point has credibility, there's a methods section. The methods section, where researchers describe the activities they completed to test their idea and the people and/or materials used in their research, is essential if someone wants to replicate the research. For articles that present the result of experimentation, whether it's a form of lab experimentation or an experiment conducted by using some other means, the methods section is one of the most important of the article.

In the next section, researchers may report the results or findings; that is, they provide the raw data that they analyzed to form the conclusions they are presenting. Finally, they discuss their work, pointing out possible shortcomings or ways to improve upon the research design. They might indicate the significance of their work and suggest ways that other researchers might branch off on their research, based on the findings of this study.

The research article therefore provides a logical pattern for readers to understand the process for conducting and reporting research:

1. Study what's going on within a subject area or a field, and what's gone on before.
2. Develop an idea that can be further researched or tested.
3. Develop an approach for testing the idea and recording data about the test.
4. Analyze the results of the test.
5. Discuss how the research could be improved if it is done again, what was particularly interesting or helpful about the results, and where any researcher might go, armed with the information from the current study.

This five-step process follows what's been called the scientific, or empirical, method. When you're developing ideas you want to research, following this kind of methodical approach helps to ensure that your methods, analyses, and conclusions are accurate and useful.

Each discipline has its own guidelines for original, scientific (empirical) research. But the purpose for empirical research is always the same: to test an idea and, by testing it and gathering new information, to add to the current knowledge about a subject. Whether you develop an idea and test your hypothesis or replicate another researcher's work to see if your results will be similar, the impetus behind experimentation is to learn more. Although many practical applications often come out of experiments, they are a nice benefit from the research—not its purpose.

In this chapter, you'll get some ideas for generally following the scientific method for conducting research. Within your own area(s) of expertise, you can add tasks or follow guidelines established within your field.

Theory, Paradigm, and Hypothesis

Our understanding of a discipline is framed by current theories, commonly used research methods, and interpretation of information gathered from research. If the research is considered to be scientific, it must follow established methods and operate within current theories. Although some people define a paradigm as a theory, the terms have distinct meanings. A *paradigm* is the framework or a model for understanding a discipline; it is made up of several theories about how and why things operate. The paradigm provides the "big picture" of the knowledge within a profession or a discipline. Theories make up a paradigm; they are not the paradigm.

A *theory* is a logical explanation of what is; it explains the ways things operate. A theory is not a philosophy indicating beliefs or values; it is an objective statement. A theory may consist of interrelated statements that explain what something does or is and why. It involves a systematic, methodical approach to building an explanation, which is why a theory is usually the culmination of several explanatory statements that lead to a concluding statement of what is.

Another term used frequently in empirical research is *hypothesis*. The word *hypothesis* indicates an unproven idea, one that must be tested to determine if the results of an experiment under certain conditions can be generalized to other situations. In simple terms, a hypothesis is a statement indicating the way things are expected to work, based logically on a theory. When you test a hypothesis, you're testing whether, in fact, your expectation of what would occur actually occurs. With one experiment, you can't prove that your idea is correct, but you can begin to build a base of information that suggests an idea has merit. Through additional testing under different conditions, if your results are similar or the same, your hypothesis gains credibility. But it may be a long time, if ever, before what was a hypothesis is a fact.

Hypotheses can be used to test a theory; based on how researchers believe something works and why it works that way, they can help build or support a theory. Conversely, hypotheses can result from interpreting research data, by drawing conclusions from experiments. A hypothesis, then, can be the impetus for research or the result of it.

Research helps to build theories by adding to the body of knowledge about the way something is. As more information is gathered and more hypotheses are developed and tested, theories may change, becoming more complex as more is understood about a subject or perhaps even being abandoned as new theories take their place. When researchers conduct empirical research, they establish methods that help them observe what is, not what they think should be. Their methods help them establish approaches to observe occurrences objectively and record findings accurately.

Study What's Going On in Your Discipline

As you've read elsewhere in this book, your research process never stops, if you want to be an effective researcher. Whether you're gathering information for a current project or just keeping up with what's going on in your profession, you need to keep reading, attending seminars, talking with other professionals, and so on. You're constantly learning about trends in hot topics within your profession and the key researchers, writers, developers, and designers who are currently working on projects. By studying technical communication and further developing your technical, scientific, and communication skills, you gain an understanding of the underpinnings of your profession, including, for example, the history of your discipline and important people who helped shape the current state of the profession. As you work as a technical communicator, you also understand how the profession keeps changing and how those changes will affect the future. And you probably have strong ideas about what you think that future should be.

During your daily work, through your reading, or by applying what you read to what you do, you may find yourself believing that the literature you've read was incomplete, or even wrong. You may believe that different methods would create different results, or you might find that you can apply the results differently than what is currently being done. You want to add information to the body of knowledge about a subject, perhaps to build on the theory of what is currently known.

One method of gathering evidence and testing a hypothesis is empirical experimentation. To conduct an experiment, you usually develop a hypothesis and a plan for testing it. Before you test the hypothesis, however, you first discover if others have tested the same or similar hypotheses, what methods they used, and what results they obtained. Then you develop your ideas and questions that can be tested during an experiment.

Gathering Background Information on Which to Base Your Experiment

Before you begin what you consider experimental research, you need to assure yourself that what you want to do is, in fact, experimental. You want to do research that will tell you whether others have tested all or part of your idea, which procedures they used to test it, and what their results were. That is, you want to gather all the background information you can. That means checking the published literature as thoroughly as possible, but it might also involve such activities as checking with granting agencies to see if similar experiments have been funded and are in process.

Reviewing the Literature

You might have several good ideas and an understanding of what is commonly known and what might be fertile areas for research. But before you begin to design an experiment, you need to know specifically who has worked on similar research and how much research has been conducted about your specific topic or budding hypothesis.

A good place to start when you want formally to conduct original research is the literature relating to your field of interest. Few ideas are completely original; most are based on someone else's ideas or modifications of commonly accepted ideas. The literature serves a number of purposes as you consider conducting an experiment. It provides a history and an analysis of research similar to that you propose. It helps you determine which parts of your work may replicate others' work. It's a history of others' successes and disappointments, a record of processes to follow and pitfalls to avoid.

To find the appropriate literature, begin with computerized searches of databases used in your field. In addition, search issues of scholarly journals for descriptions of previous or continuing research, read dissertation and thesis abstracts, and peruse conference proceedings and programs for indications of the researchers involved with current projects and the scope of the research.

After you have an overview of the amount and kind of information about your topic, study in-depth reports of experiments that directly pertain to your work. These reports and descriptions form the basis of your literature review, because your work either may replicate someone else's experiment or be a modification or a furthering of someone else's work. You should understand the similarities between your hypothesis and another researcher's, as well as compare the research methodologies.

By learning what has worked well, which information-gathering tools have been proven valid, as well as what has not gone as anticipated, you save a lot of time by not "reinventing the wheel." Your objective as a researcher is not to redo someone else's work, but to replicate an experiment or develop a more effective way to test your hypothesis.

Developing a Hypothesis

When you design an experiment, you frequently test a hypothesis, which may have been derived from previous analyses of experimental data or may be the impetus for conducting an experiment. For example, your hypothesis may be derived from other researchers' experiments; you may change one or more variables and anticipate what will happen. You write the hypothesis and determine how you can test it.

You state your hypothesis in clear terms that spell out the purpose of the research. *If-then* statements are often useful in hypothesis development. *If* a certain event or phenomenon takes place under these conditions, *then* the event or phenomenon is expected to occur under these conditions.

This cause-and-effect type of statement is only one type of hypothesis, and you want to be sure that you and anyone else working with you understands that seldom does one effect result from only one cause, or that one cause leads to only one effect. You anticipate what will happen, but only by replicating the experiment several times or changing the variables and developing new experiments can you be more nearly certain that A leads to B, or that B is the result of A. Often, you determine the likelihood of a cause-and-effect relationship, but chance and other unknowns can affect the outcome of what you're predicting. As you develop methods for conducting the experiment, you also devise ways to limit things occurring purely because of chance or find ways to limit the possibility of error being introduced.

Another hypothesis statement might be something like, "At the three Eastern universities used in this survey, students under 25 years of age are more likely than nontraditional students 25 years or older to use electronic media more than textbooks as their primary source of course information." This statement is specific, in that it indicates the parameters of the research: information provided by students at three Eastern universities, divided by age, who responded to a survey. The hypothesis you're developing can be tested once or several times to learn how the respondents supported or negated your idea about the primary sources of course information. But before you generalize this hypothesis to every university in any country, you need a great deal more supporting data.

The hypothesis you develop should be testable. It should be as logical as you can make it without further testing. It should have meaning, so that the results you get will be useful in providing new data about a subject. It should be clearly stated so that it serves as the catalyst for the experiment you design. Once you've developed a hypothesis, you determine where, when, and under what kind of conditions the experiment should take place.

Experimental Conditions

You must make several decisions as you prepare your experiment. You determine which variables should be compared, which conditions should be present during the experiment, and how you can best conduct the experiment.

As you set up an experiment, you look at variables and attributes. An *attribute* is a value, characteristic, or quality that describes someone or something. For example, attributes of a specific student may be male, Asian, Democrat, and technical communicator. A *variable* is a grouping of attributes. Gender is the variable that groups male and female, for example, and technical communicator is part of a variable for occupation. As you plan an experiment, you most often analyze the relationship between or among variables, such as gender and occupation, or ethnicity and socioeconomic level, or age and educational level. In the survey example, the hypothesis indicates that age and use of electronic media are the variables being compared in this research.

You might also determine which variables are givens, or *independent variables,* such as age. Independent variables can be manipulated so that you can learn the effects of one variable on another, the dependent variable. You might, for example, see if different groups of people of different ages react differently to a stimulus. The reaction may change, depending upon the age group being tested

Other variables may be *control variables,* or constant variables, such as using gender as a constant basis of comparison against other variables. Comparing men and women across other variables, such as the type of response they gave to a question, their economic status, or their job title, for example, could involve the control variable for gender.

Dependent variables may vary or be caused by the independent variable, ones that depend on the independent variable. It is the variable that's influenced by varying the independent variables. For example, if age is an independent variable, you vary the age groups of people responding to a question, the response to the question is the dependent variable. The response is a dependent, studyable variable; everyone gives a response, which you later analyze. The responses may vary, depending upon the age of the person answering the question.

You might also use *moderator variables*. These variables provide a context for the experiment. They represent existing characteristics or qualities of the people or materials being studied. Moderator variables might be previous training, completion of an academic degree program, or marital status. People who participate in your research may have characteristics such as having taken desktop publishing training in the past two weeks, being graduated from a four-year college program with a B.A. degree, and being married within the past year. (Additional definitions of variables are provided in Chapter 13, Statistics and Measurements.)

Although a thorough discussion of ways to design experiments, including the importance of determining the relationships between and among participants or materials, comparing variables, and determining possible errors because of chance or methodological flaws, is beyond the scope of one chapter, you should at least begin to have an idea of the complexity involved in planning empirical research. You may need to take an additional course in research methodologies specific to your discipline or make your designs reflect practices commonly followed within your discipline.

As general guidelines, the following questions should help you analyze the best way to test your hypothesis:

- Where can you best test the hypothesis (such as in a lab, in an office, in a field, in a classroom, on the street, within a home)?
- Under what conditions should the experiment take place (such as in a sterile or controlled environment, in a "normal" setting, during a natural event like a thunderstorm)?
- What kind of control variables do you need?
- What kind of variables should you manipulate?
- When should the experiment take place?
- How long should the experiment be?
- What kind of materials will you need to conduct the experiment (such as beakers, protective eyewear, sandpaper, questionnaire respondents, a valid survey)?

- If you need to work with people or animals in the experiment, how many participants are necessary to make an effective group? Where can you get such a group?
- What kind of equipment will you need to conduct the experiment (such as anemometers, thermometers, computers)?
- Who will conduct the experiment?
- How will you/they conduct the experiment?
- How will you/they collect data?

The answers to these and related questions help you develop a research design. For example, if your hypothesis is "At the three Eastern universities used in this survey, students under 25 years of age are more likely than nontraditional students 25 years or older to use electronic media more than textbooks as their primary source of course information," you already know a great deal about the design of your experiment.

You might find it useful to break down your hypothesis into important words or phrases. For example, the first phrase you need to deal with is "three Eastern universities." This phrase can generate several questions that you'll have to answer as you develop your research design.

Where do you want to conduct the research? Obviously, you have in mind three Eastern universities. But which ones? Do you want institutions that are similar in the makeup of the student populations, or ones that are different? Do you want institutions that are private or public, rural or urban, large or small, with specialized programs or those with a general curriculum? Why are Eastern universities important? Do you have an idea that students in the Eastern United States act differently than students in other regions? Why? Did you choose Eastern universities because you or close colleagues work in Eastern universities? If that's true, are you as objective as you can be about the research? Or will your familiarity help you gain access to a group of students you'd like to study? You need to narrow your response to the where question by determining which institutions will be the best setting at which to test your hypothesis. From the responses to these follow-up questions, you can narrow your research site to a list of appropriate institutions. Then you'll have to gain approval to work with students, and probably faculty and administrators, too, as you distribute and collect the surveys.

Even when you've targeted specific universities, you need to refine your plan. Where exactly do you plan to survey students? Are you going to solicit their help in an open setting, such as the campus bookstore, lounges, dining halls, snack bars, or an open commons area? Would your hypothesis be better tested with a more controlled group, such as a class? If so, which classes should be targeted? Do you want students with a particular major and minor, or do you want students from a general class that is taken by students from a variety of majors?

The next important word is *survey.* A survey may be shorter than a questionnaire. Is that your intent? How many questions do you envision? Who'll create this survey? Will it be based on a survey that someone else developed, one that has been validated and its use replicated through previous studies? If you're developing a survey, what do you want to ask? How do you plan to ask the questions? How will people answer? How will the survey be administered? How will copies be collected? If you're creating the survey, how will you validate it before you base your experiment on its use?

"Students under 25 years of age" is the next important phrase, and again, you need to answer a number of questions as you develop your methodology. How many students do you need? Should they be in one class, or should they represent all of the student body, from different classes? How will you know they are under 25? Will you ask them? Will they fill out a demographic item on the survey? Will you check school records? Why is 25 the cutoff age? You can ask similar questions about the next group of participants indicated in your hypothesis: nontraditional students 25 years or older. What makes these students nontraditional? How do you define that term? How much older is older? Do you need a cutoff upper limit, or will you just report the upper age limit from the group of respondents?

Now you might consider defining and describing what you mean by the rest of the hypothesis: that younger students "are more likely to use electronic media more than textbooks as their primary source of course information." What do you mean by "are more likely"? Do they use electronic media more often, in more situations, in different ways, because they have more access, because they enjoy working with electronic media more than reading textbooks? You may decide that you need to restate this part of the hypothesis. And why did you decide only to study *if* they are more likely to use electronic media more than textbooks? You might study *why they say* they are more likely to do so, or *how much more likely* they are to use electronic media, or *why they do use* electronic media more than textbooks. Again, you may decide to alter the emphasis of your research, if you decide that you want to study more than what you originally indicated you would study.

Also, you know as you look at the rest of the hypothesis that you'll need to provide some operational definitions for your colleagues who might help you with the research or might eventually read about your methods and findings. Even more important, you need to specify how you're defining terms in your survey, so that your respondents will work from the same definition you do. For example, how do you define electronic media? Do you plan to include e-mail, teleconferencing, use of the Internet, use of other networks and databases, bulletin boards, interactive CD-ROM, software, and so on (which you will need to list as part of your definition)? Are you simply indicating general computer use? Is videotaped information played on a personal VCR and television programs considered part of electronic media? The term as it's listed in the hypothesis can generate many different meanings among participants and other researchers, so you need to provide a clear definition.

Another term that should be defined in your research is textbooks. That may seem like an obvious definition, but some students may perceive a textbook differently than you do. In some classes, a textbook may be a traditional textbook—a bound book containing information students need to know while they study a particular subject. But a textbook in other classes might be a photocopy of the professor's notes, which can be purchased in the bookstore or from a print shop. It might be interactive software that's been customized for the class. It might require students to have a computerized class account in the campus computer system so that information can be uploaded and downloaded between professor and student. It might be audiotapes, videotapes, samples, workbooks, or a variety of other sources of course information. And some professors may not use a textbook at all, preferring to lecture, provide workshops, or lead field site visits. You may need to list these options, or limit your definition to a traditional concept of a textbook.

Two other phrases that need clarification are "primary source" and "course information." What do you mean by a primary source? Is it the way students are most often *expected* to gather information, or is it the one they *most often use* to gather information? Can there be only one primary source of course information? And what do you mean by "course information"? Is this the information students formally are supposed to learn throughout the term—the subject matter studied within a course? Or does it include information students learn about the course—how it's designed, which procedures are followed, what other students think about the course, how the professor is perceived, when assignments are due—in short, information about the course itself, not the subject matter.

So, even when you answer the general questions about your research, you ask a number of additional questions to refine and make specific your answers. And each answer should be a logical way to test your hypothesis. If you, as many researchers do, discover that your hypothesis is too specific or too general to be tested, you first need to return to the hypothesis and develop a better statement of your research idea.

Establishing a Procedure

In addition to these questions relating to the experiment itself, you need to consider the practical matters of getting approval for your research methodology and receiving the time, materials, subjects, equipment, and space you need for the experiment. These questions help you analyze the managerial aspects of your research:

- How will you get the materials, equipment, and participants for the experiment?
- How will you get approval for the research from funding or regulatory agencies?
- How will you get approval for the research from ethics boards, such as a human subjects review board?
- How will you fund the experiment?
- What kinds of procedures will you establish to ensure objectivity in the research or reduce bias to acceptable levels?

Answers to these and related follow-up questions can help you devise a good method. Then, during the experiment, you follow that method. After you answer all these questions and determine ways to gather the materials, equipment, and participants and gain approval to conduct the research, you can set up the experiment.

Creating Forms, Documentation, and Tools

Before you conduct your experiment, you probably need to create and possibly test several documents that will be used during the experiment or to document your work. For example, you may create an informed consent form and have the form approved before you can solicit help from people who'll take part in your experiment. You may need lab report forms, copies of a questionnaire, or copies of a test.

Before you ever let a participant or another researcher use your documents, you should have them tested to ensure their validity. You might decide to use a research instrument, such as a test, a questionnaire, or a form, that has been verified for accuracy and tested before for reliability. When you use someone else's instrument, such as an established test or questionnaire, you may need to pay a fee for copies ordered from a publisher or you may need to get copyright permission to use the research tool. If you develop your own research instrument, you submit it to several practice tests to ensure that it's the most effective, reliable tool possible for this experiment.

In the survey example, you would get approval from the universities' human subjects review board, because your work involved the participation of people. After you filled out forms and perhaps met with the board, you would describe your research methods and indicate the ways you would gather information safely and confidentially from students. When the research has been approved, you then would develop the appropriate forms to describe the survey for students and to gain their informed consent to participate in the experiment. You would also need an adequate number of copies of the survey. You might have developed special instruction sheets for the colleagues who help you administer the survey. Although you might possibly create even more forms, these materials would be critical to the success of your experiment.

Selecting Participants

Participants may be an important ingredient in your experiment, and you should select them carefully. There are a number of ways to select the people who will be involved in your experiment, and you might consult an expert in research design or even a statistician to determine the best sampling method for you to use for your research. You might use probability or nonprobability samples in your research. In a probability sampling method, such as a simple random sampling, each person (or animal, species, and so on) has an equal chance of being selected. In a nonprobability sampling method, such as a judgmental sampling, a person may have a greater or a lesser chance than someone else of being selected. Of course, you would need to consult with experts who routinely conduct these sampling methods to learn more about which one(s) you should consider for your experiment.

In the survey example, you might decide to use a convenience sample of students in classes your colleagues or you teach, simply because these students are available. However, in the limitations of your study, you should note that the participants were part of a convenience sample and may not be representative of students at other institutions or of the general population of adults.

Table 11.1 provides you with a general idea of some ways to select participants.

Table 11.1 Some Sampling Techniques

Cluster Sampling	A cluster, or subgroup of participants, is selected for sampling because the entire group would be impossible to identify. For example, a list of all students in the United States might be difficult, at best, to create, but students listed by university would be possible. The university population is a cluster, or subgroup, of the larger student population of the United States.
Convenience Sampling	Participants are selected because they are a convenient group. For example, students on a college campus may be selected to participate in professors' experiments because they are a convenient group to work with.
Purposive Sampling	The researcher selects participants for the study, using a nonprobability sampling method. This is a judgmental sample. The researcher "judges" who should be in the sample group.
Quota Sampling	A matrix describing the characteristics of the target population and the relative proportion of people with each characteristic. Each cell, which includes the number of people with a certain characteristic or set of characteristics, can be sampled.
Random Sampling (Simple)	All participants have an equal chance of being selected for the experiment. Some simple ways to do this include drawing a number from a hat or being selected by lottery.
Snowball Sampling	The experiment begins by using a few participants who have been selected. Other participants are gained through referrals from the original participants. Each group of participants can provide additional referrals.
Stratified Sampling	All members of a population are grouped into homogeneous strata before the population is sampled. For example, a strata might be ethnicity, race, sex, or socioeconomic level. Within the strata, members are randomly sampled.
Systematic Sampling	A more elaborate form of simple random sampling, in systematic sampling, every *n*th item from a list is selected.

Training and Special Activities

If other researchers, either volunteers or colleagues, are involved in gathering data, you may set up training sessions so that everyone is familiar with the methods you'll use and will follow the procedures exactly. For example, if several researchers will be helping you distribute surveys to students during their classes, each researcher should know exactly what to say and do when working with the professor, when instructing students how to complete and return the survey, and when the surveys have been collected. You may provide a script to read, so that all respondents receive the same information, for example. Or you may conduct practice sessions in which each researcher's style, language, message, and presentation are critiqued for consistency and effectiveness.

Training sessions, a run-through of the experiment, and any other special activities that help you and your colleagues understand thoroughly how to conduct the experiment are a good investment in time. When everyone is familiar with his or her responsibilities and understands exactly how to follow procedures, the quality of the research should be high.

Setting Up Materials and Equipment

Before the experiment, double check your materials and equipment to make sure that you have everything you need and that everything is working properly. If you are setting up a conference room, for example, for interviews with participants, check the room to make sure you've created the appropriate atmosphere. Just a final check can help to ensure that you're ready to conduct the experiment.

During the Experiment

During the experiment, you—and any colleagues involved in the data collection—keep an open mind. After all, the purpose of your research is to see what happens, not to structure the experiment to prove your idea. Some of the best research findings are gathered just because the results were unexpected and helped researchers find new connections between concepts and variables. Being as objective as possible is a plus.

You also need to be methodical. If you've planned procedures to follow and recording practices that ensure consistency in describing observations and recording data, then you'll be able to logically analyze your findings and draw conclusions from the results you receive. Once you've established a procedure, you and your colleagues stick to that procedure.

Interpretation of Results

When researchers record the results of experiments, they must scrupulously keep records of observations and data received from people, animals, and/or equipment. As a researcher, you keep track of the numbers of participants, amounts of materials, times, temperatures, responses to stimuli, and any number of other variables that may be operating in your experiment. You account for every moment of the experiment and every condition under which the experiment took place. Your methodology must be consistent and verified through your and other researchers' notes and records. In this way, you can trust your results and know that as closely as possible you limited any unaccounted variables from straying into your experiment.

You should have someone help you interpret results to ensure the objectivity of your methods and to check the logic of your analysis. Good researchers understand that generalizing from a small sample of results or one experiment is a dangerous practice. Therefore, you cautiously interpret the results of your experiment and plan to replicate, or have someone else replicate, the experiment. The more consistent the results from numerous experiments, the more likely that your hypothesis will be proven. A good statistician can help you analyze numerical data and provide you with an analysis of the results. (You'll learn more about some statistical terms in Chapter 13, Statistics and Measurements.) Other researchers can help you interpret your findings and see if they can be generalized to other settings and situations.

Reports of Your Findings

Researchers conduct experiments to share their findings with others. Therefore, if you design experiments and test hypotheses, at some point you expect to share your findings with others. You may be asked to present a paper at a conference of your peers, write an article for a scholarly journal, delineate your work in a book, or at the very least write a report justifying your recent work for a class or company. You should get in the habit of outlining the research methodology before you begin the experiment, making sure your hypothesis is sound; carefully conducting the experiment; logically analyzing the findings; and systematically evaluating the need for additional hypothesis testing. If you keep careful records and have thoroughly accounted for every part of the testing process, you have the makings of a good researcher, one whose findings can advance our understanding of a subject.

Summary

A first step in conducting empirical research is learning more about your discipline. Literature reviews, not only reflecting your specific research interests but a general interest in what's going on in your discipline, can help you learn more about the theories upon which your discipline's paradigm is based. The more you understand about the principles and purposes of experimental research, the more you can apply what you know to your reading of the literature and your understanding of others' research.

Hypothesis development and testing are important parts of research, but you must be careful to be as objective and methodical as possible when you devise an experiment and conduct it. Although your discipline or profession may have additional guidelines or standards for conducting experiments, you should at least follow the steps and use the questions in this chapter to help you develop a good research plan.

You should also check with all regulatory agencies that might in any way need to approve your research, because many institutions and federal agencies that sponsor research have lengthy procedures to ensure the safety of participants and the validity of the research methodology.

Research Activities

1. Write your plan for an experiment, including the literature review that provided you with the idea for your research. Then have the appropriate authorities approve your work.
2. Read the Call for Papers section of several scholarly journals to find a topic for possible experimentation.

3. Read a recent edition of *Dissertation Abstracts* to learn the type of research going on in your field.
4. Get a copy of the guidelines for experimental research conducted by your institution.
5. Replicate another researcher's experiment, perhaps as outlined in a textbook or a journal. Practice following an established research method and match your findings against those of the original researcher's.

For Further Reading

These are only a few of the many good books about experimental methods in some of the many disciplines that make up technical communication. They are representative of other books, including textbooks, that you might want to consult in your scientific or technical specialization.

Babbie, Earl. 1995. *The Practice of Social Research.* 7th ed. Belmont, CA: Wadsworth.

Halpern, Jeanne W. 1988. "Getting in Deep: Using Qualitative Research in Business and Technical Communication." *Journal of Business and Technical Communication* (September): 22-43.

McGuigan, Frank J. 1993. *Experimental Psychology: Methods of Research.* 6th ed. Englewood Cliffs, NJ: Prentice Hall.

Stacks, Don W., and John E. Hocking. 1992. *Essentials of Communication Research.* New York: HarperCollins.

Zikmund, William G. 1991. *Business Research Methods.* 3rd ed. Chicago: Dryden Press.

PART THREE

Ways to Evaluate Research

Research is a process of evaluation and revision. Throughout, from the time we get an assignment or have an idea we want to pursue until we've published and received feedback, we're involved with evaluating our research. When we have an idea, we evaluate and revise it until we have it focused. When we have an assignment, we plan it, then evaluate and revise the plan. We look at research methodologies, plan them, evaluate them in terms of the quality of information we'll get for a particular project, and revise them.

As we discover sources, we make preliminary evaluations; then we use them and evaluate them again. Some sources of information and research methods may have been extremely helpful, and you know exactly what you want to write. Or you may have conducted experimental research and have what appears to be a wealth of raw data, all needing to be evaluated and interpreted before it can be shared with others. Or you may find that your research leaves you somewhere in between: You have some information that you can use when you write a document for your audience, but you also have some conflicting information. In any of these typical research situations, you have to evaluate the information you have, determine if you need to go back and gather more information, or determine how to reconcile conflicting sources. The chapters in Part III will help you work with the data you've accumulated.

Part III consists of these chapters:

- Chapter 12, Evaluation of Sources
- Chapter 13, Statistics and Measurements
- Chapter 14, Usability Testing and Validation

Chapter 12 helps you reconcile conflicting sources of information and compare the usability of different sources. In Chapter 13, you learn some basic terminology used in statistics and read about some tests that you—or more likely, a statistical expert with whom you consult—can conduct with the data you've gathered. Although in one chapter you won't learn everything about statistical methods, you can develop a vocabulary to use when you work with statistical experts or as you further develop your skills in working with statistical information. Finally, Chapter 14 provides a practical overview of conducting usability tests and validating documents you've created, based on your research. The information in Chapter 14 can be used to help you test a document (or another product) you've created, or to conduct and incorporate usability tests as part of your company's or your personal ongoing research.

CHAPTER TWELVE

Evaluation of Sources

So far, you've seen how you gather information from printed and online sources; primary sources like experiments, interviews, questionnaires, letters of inquiry, and personal experience; secondary sources; and possibly tertiary sources. But what does a good researcher do when the sources may conflict with one another? Especially in technical communication, finding the most reliable and accurate source is crucial, perhaps to the safety, and definitely for the convenience of readers. However, as you gather information, you often find that sources, even highly reputable sources, may disagree about important information, or may interpret results in different ways. And, as experts may have different perspectives about or experiences with a product or a process, they may have equally good—but differing—approaches to doing something or to information about a topic. How do you get a good idea of the complete picture? How can you reconcile all the bits of information you gather so that what you ultimately describe to an audience is true and as complete as possible?

As a technical communicator, you may be called upon to make sense out of this conflicting research and to present the best form of information you have gathered. How can you make that decision? This chapter provides you with guidance to help you decide which pieces of information best fit together for your audience.

An Information-management Strategy

To sift the information and determine the best sources, whether you are compiling documentation for a new product, describing a process or developing instructions, analyzing a scientific problem, or completing other research tasks so that you can write a document, you evaluate these four areas:

- the reliability of the source
- the age of the information
- the quality of the information
- the style or format of the source

Depending upon the type of source, you might find that it scores high on more than one area of comparison. Some sources may be reliable, current, high-quality, and effectively presented—in which case, you're very lucky, and your information may almost write itself. More likely, however, is that your evaluation indicates some strengths and weaknesses in every source you use. Then you need to compare sources among these four areas to see which can be used most effectively in your document.

For example, some sources, such as a subject matter expert with whom you've worked for years, may be a highly reputable source, but the way in which he or she presented you with information was not as useful as it could have been. Perhaps you received a faxed page of handwritten notes that were barely legible. Although the information was good, the style or format wasn't very helpful. Nevertheless, you saw some surprising information in the notes, which you hadn't found or heard anywhere else. You also found a thorough description of the subject in a secondary source—a journal article—but the information was more than six months old and probably outdated because it dealt with a rapidly changing subject. The expert provided a new insight, but not enough information for you to write a complete document. The article provided a greater depth of information, but it may be outdated. Of these two sources, how can you get enough useful information? Where can you go next?

You might brainstorm, freewrite, or complete any other idea-generation activity to determine what you know and what you still need to know, especially about your expert's notes. Then you might be able to set up an informational interview with the expert, or at least write or call to request clarification of the notes or answers to specific questions. You might complete an online search of recent research articles to learn if newer articles have been written about the topic. As you receive more information, you add the pieces to your puzzle to help complete the picture of this topic. If you can't gather enough new information in this way, you can search for additional sources to fill in the gaps left by the sources you already have. Ultimately, you may find that the information from the expert may provide insight into new directions about the subject, and the journal article may be useful as background information to remind your audience of what has recently occurred.

If your subject was *laser surgery*, and you were assigned to write a newsletter article describing what's new in this field, you might have surveyed literature from the past six months to provide an overview of what your audience may already know about laser surgery or what took place recently. Then, as you described new techniques in laser surgery, you might have effectively included quotations from your surgical expert and indicated how laser surgery might be used in the near future. The information you originally gathered was useful in some ways, but neither source provided a complete picture of laser surgery in the form that your audience needed. With additional sources, you filled in the gaps in the information and presented your audience with an accurate view of the complete subject, at least as of the newsletter's publication date.

At other times, you may be able to find only two sources, such as experts, who present two very different views of the same subject. For example, before you could write training materials for shopfloor employees, you interviewed a shopfloor supervisor and an employee with many years of seniority who works with the equipment daily. You asked each, separately, about the best way to operate a piece of equipment. The supervisor presented you with complete information, including a demonstration, about a safe, efficient method for operating the equipment. The technician who operates the equipment every day showed you a number of safe, effective shortcuts to the job. The technician's process seemed easier, but the supervisor knew how the job was supposed to be completed. Which method should you promote in your training materials?

In this situation, you might find that both sources are reputable and reliable. Their information is current and reflects safe practices. The quality of the information is high—both people provided you with complete, validated information. The form in which they presented the information—a demonstration—was effective and helped you to understand that both approaches to operating the equipment would be useful in training. But you can only show one way to operate the equipment, because everyone should be trained to do the job in the same way. Although it is a more difficult method, ultimately you may need to use the supervisor's approach, because the supervisor has more authority on the shopfloor. However, you might indicate some shortcuts for some steps in a "helpful tips" section, for example.

By comparing your evaluations of each source's information in all four areas, you should be able to pick the most reliable information from all your research findings, or develop enough questions and ideas to conduct additional research. The sources you've already located may then be replaced or supported, based on new information.

Reliability of the Source

The source's reputation or reliability is the first item you should check, usually even before you determine which sources you'll consult. For example, when the research project was first assigned to you, or when you came up with an original hypothesis for testing, a number of sources competed for your attention. At that time, you narrowed the number of possible sources by deciding which people should be interviewed, which printed sources should be consulted for the best background information, and so on. Then, by the time the research findings were ready to be compiled and the report, article, or study written, you had a good idea about the validity of the sources.

At this point in the research, however, you may have noted some discrepancies among sources. For example, you're responsible for writing a report about the effects on personnel of buying some new computer equipment. Two people you interviewed said that buying five new PCs and adding them to the network would force the company to fire two employees, who would no longer be needed for typing manuscripts and proofreading copy. But three people believed that the new computers would create a new position for a training specialist, who would be responsible for retraining employees to work with computers and follow new procedures within the department. Which interviewees should you cite in your report? All are credible sources, but their analyses of the situation differ. That's when you need to evaluate their reputations and reliability.

You found that two people have been accurate in assessing the impact of previously purchased equipment. They've been with the company longer than the other employees, and they've seen how the equipment versus people issues are usually resolved within the department. These people have a reputation for accuracy, and they've been reliable sources in the past. Another person worked as a typist for a year before being promoted into a new position and is very loyal to his co-workers. Nevertheless, he believes that the typist position within the company is becoming outdated and probably will be replaced within a year, regardless if the equipment is purchased. The other people you interviewed may not have as strong a background of experience or personal knowledge of the effects of equipment purchases on personnel. Although all your sources are credible, some are more reliable in this situation.

At times, you might have to defer to someone's authority, or power, in the workplace. When two sources seem equally reliable, you may have to select one source instead of another based on his or her position of power. For example, a manager has more authority, and presumably knows more about the company as a whole, than an editor working in the publications department. However, that editor has more authority than data entry specialists who work in another area of the company. But the person with the most authority may be the CEO who has to make the final decisions about personnel and procedures at the company.

Reliability and a *good reputation* can mean many different things, depending on the source. A person's reputation and reliability as a source may come from familiarity with a subject, on-the-job responsibilities and experiences, education, interests, and degree of professional development. A good reputation may indicate a person who is as free from prejudice as possible, someone who will not mislead an audience, manipulate information for personal gain, or suppress information for any reason. A reliable source is someone who knows a subject well, has responsibility for making decisions based on that knowledge and experience, may publish research or comments on the current nature of a specific field, and can be counted on to provide all the facts he or she knows. You may not find only one source who becomes *the* authority on the subject you are researching, but you can distinguish among the people who know the subject well and have recent firsthand experience with it. You can determine who may not be as familiar with the subject, and who, therefore, should not be given the most credibility when you are faced with conflicting pieces of information from your sources.

Reliability in questionnaires and surveys, for example, can come from consensus, often among several tests or uses of the same survey tool. If the majority of respondents answered questions in a certain way, or provided similar comments to open-ended questions, that questionnaire

Questions about an Author's Reliability

- What are the author's credentials?
- What are the author's educational background, employment background, chief interests, publications, and professional activities?
- Is the author speaking in his or her field of expertise?
- Is the author up to date about the subject?
- Have the author's other publications been reliable?
- Have the author's data been personally obtained or used secondhand?
- Does the author appear knowledgeable about the subject?
- Has the author clearly separated interpretation from fact?
- Has the author used reputable sources in the research, as listed in the bibliography?
- Might the author be susceptible to pressures (personal, institutional, or political) that would cause him or her to alter or suppress information?
- Is the author's experimental sample reliable or verifiable?
- Does the author gloss over contradictory evidence?
- Are the author's findings or conclusions based on an exaggeration of the facts?

Questions about Publishers and Their Publications

- What is the reliability of other documents it publishes?
- Is this a publisher or a publication cited by leading researchers?
- Do leading researchers submit work to this publication?
- Who are the editors and editorial board, and what are their credentials?
- Are documents reviewed by experts before publication?
- Does a quick examination of the document's table of contents indicate whether the subjects are adequately covered?

Figure 12.1 Questions to evaluate reliability.

may be reliable. It may be reusable in other similar situations, because respondents understand the questions and seem to provide responses accurately.

Secondary sources of information are very reliable, too, although in using secondary sources, you work with published information instead of people. Reliability comes from the author's, publisher's, or publication's reputation.

Just as you evaluate an interviewee's credentials, so can you assess an author's credentials based on his or her education, prior publications, degree of familiarity with the subject, and job title or work experiences. Authors known for high-quality publications and/or numerous publications within a profession can be counted on to provide you with solid research, which you, in turn, can use in your research.

A publisher's reliability rests with the type of works published by the company. Noted publishing houses develop a reputation based on the types of books they most often publish. A reliable publisher checks the author's credentials and the content of the work before it is published. Although smaller publishing houses may not be as well known, they can be equally reliable. They may specialize in one type of publication, such as business communication, or they may be associated with a university press. The quality of a publishing company's work is more important than the company's size. To check a publisher's credentials, if you don't know the publishing company's reputation, you might check reference books like *Writer's Market* or *Fortune 500's Directory of Businesses*. Reference works like these often contain a description of the publishing company's holdings and their circulation. *Books in Print* also contains information about publishers, as well as the works they've recently published.

Journals, like other publications, can be evaluated for their reliability, too. Journals associated with a particular discipline generally have a better reputation for dealing with their specialized subject matter than do general magazines. Journals with regularly published articles in specializations like microbiology, physical sciences, mathematics, training, business management, and management information systems, for example, usually provide more in-depth information and describe current research in that specialty more often than popular magazines for a general audience, such as *Time, Newsweek,* and *U.S. News and World Report.* Although these magazines are useful and provide valuable overviews of a number of subject areas, they do not provide the kind of information that experts working within one specialization need as references. For ideas on the type of research being done in technical communication or topics of current interest within the profession of technical communication, for example, one reliable journal is *Technical Communication.* Teachers who need research ideas involving student subjects or who want to see how their research can be applied to the education of future technical communicators might turn to journals like *Technical Communication Quarterly* or *The Journal of Business and Technical Communication.*

Because you have special research interests in a specific discipline, you should become aware of the reliable sources, whether authors, publishers, or publications, for your area of specialization. Knowing about these sources can help you determine which sources have a history of reliability and a reputation for thoroughness and fairness. Figure 12.1 shows a checklist that should help you evaluate a source's reliability and reputation.

How can you find the answers to all these questions? Most publications carry a short biography of the author or researcher, a list of editors, and publication criteria. By reading a published source carefully, you find answers to many questions. Many researchers and authors publish regularly, and with a quick review of their other works, you can often evaluate the quality of the information they usually produce. You might also find reviews, commentaries, or evaluations of the researcher's previous work, which can help you determine how highly other professionals value the work.

In the same way that people build credentials that can be checked, publications build credentials. By checking with professionals and professional societies, you can learn which publications are the most respected in a field. Doing some background research about your sources can satisfy you about their credentials and reputation. Then, if you answer these evaluation questions to your satisfaction, you have determined that your source is reliable.

Timeliness of the Information

In addition to a source's reliability or reputation, the timeliness (or publication date, interview date, or survey date) of the information is important, especially in technical communication, because technical and scientific subjects change rapidly. Thus, you should always check the copyright date of the information, if it has been presented or published, and the date of the experiment or other research if the information has not been published. Also be sure to check the publication lag time, because although the publication's date may be current, the research being reported may be several months or years old.

However, even in technical communication, newer is not always better. The information resulting from one scientist's research may still be valid, or the findings may have become scientific fact after a hypothesis has been tested numerous times with the same results. Some information may be relatively timeless, especially as a background piece or a landmark study on which additional hypotheses have been based and tested.

Some information, on the other hand, may become outdated rapidly. The user's manual for last year's product may need to be updated, for example; in fact, that may be the reason you're looking for new information. The publication date on that manual may indicate the source is not as valuable as the new information recorded from last week's interview with the product developer. In a similar way, last week's interview may be outdated, because the product developer modified the product just yesterday, and the information he or she gave you last week no longer applies to the product. A new interview should supersede last week's interview.

You may have to work harder in assessing the importance of the information's timeliness than in evaluating any other aspect of the research. If your findings conflict, the date of the information should be checked to determine if new facts have been uncovered or if the product or process has been modified recently. You should know your subject so well that you can tell if some information is timeless, or if all information needs to be updated daily. Only then can you determine which piece of information is accurate and should be used in your work.

Quality of the Information

At times, you may be faced with an overwhelming amount of research with consistent findings. The interviewees all agree, the questionnaires provide similar kinds of information, the tests produce the same results, and the publications present similar information. You might think that this would be paradise, for you often have to sift through pages or screens of research notes to create only one short document. However, too much information can be as confusing as too little or conflicting information. In this case, you have to evaluate the quality of the information. To do this, ask these questions:

- Is the information accurate?
- Is the information consistent with other researchers' findings?
- Does the information provide a different perspective on the research problem or project?
- Is the source particularly noteworthy in itself?

If a piece of information from one source doesn't stand out from information from other sources, it might be condensed or relegated to a footnote. If, on the other hand, the information is not only accurate but provides a different view to the problem, that source should be given a higher priority. For example, if you have interviewed 50 employees of a manufacturing company to find out how they feel about the hazardous material-removal process in their company, all 50 may agree that the material needs to be removed quickly and may express concern for their health. But you may get different perspectives on the problem when you review the interviews with the company president, a secretary in the front office, a supervisor working the midnight shift, a forklift operator, and the company's nurse. The interviewees agree on the severity of the problem and approve the company's procedures for removing a hazardous substance, but each interviewee has his or her own perspective on the problem, based on personal experience and knowledge of the problem. In this situation, the quality of the information is high from all sources, and you might highlight each person's perspective in a different section of your document.

Sometimes *quality* may refer to the amount of information. A source that provides a great deal of useful information may be assessed as a better source than one that provides accurate but limited information. Quality is a good way of evaluating information to be used in a final publication, but it should be evaluated only after the source's reliability and age have been evaluated.

Style or Format of the Information

As a last point of evaluation, the style or format of the information should be assessed. For instance, one source may provide information in a much more interesting, entertaining, explanatory, or illustrative way than other sources that give the same facts. A good quotation, a high-quality photograph, a descriptive observation, or a clear report, for example, may provide information in a better style or format than any other source. And a high-quality photograph may be more useful than a series of quotations that would appear in prose. The form, or medium of presentation, can be an important part of style and format, too.

Of course, a source using an interesting style but providing questionable information should not be favored over an accurate but boring source. In the case of two or more sources providing the same quality, accuracy, and reliability, you may choose to quote or include examples from the source(s) with the best style of presenting the information.

Summary

These four areas of evaluation can help you sift the information you have gathered in your research and begin to put the information in a useful form, whether you need to write a report, script, journal article, or other document. Information gathering is an important focus of research, but a good researcher needs to be able to evaluate the results of that information gathering and to choose the best way to present technical information. The researcher/writer is often a filter of information from sources to the public. With such a responsibility, you have to create methods of systematically evaluating conflicting sources or multiple sources and integrating them into publication-quality information.

Research Activities

1. Evaluate each of your sources by using the checklists provided in this section. Then rank your sources in order of most important to least important. If you've found some "bad" sources, eliminate them and return to the planning stage to gather information from better sources.
2. Do some background research on journals published in your area(s) of specialization and determine which journals are the most respected in your field. Then list those journals for future reference.

For Further Reading

Borg, Walter R., Joyce P. Gall, and Meredith D. Gall. 1993. *Applying Educational Research: A Practical Guide.* 3rd ed. New York: Longman.

Di Vittorio, Martha Montes. 1994. "Evaluating Sources of U.S. Company Data." *Database* (August): 39-44.

Losee, Robert M., Jr., and Karen A. Worley. 1993. *Research Methods 101: Research and Evaluation for Information Professionals.* San Diego: Academic Press.

CHAPTER THIRTEEN

Statistics and Measurements

One way to make sense of the data you've accumulated from experimental research, such as the responses to your questionnaire, is to analyze your findings statistically. Depending upon the type of research you've conducted, and therefore the type of data you've gathered, you might conduct one or several statistical tests. These tests can help you determine if some findings were the result of chance. They can also help you determine if some findings are significant, that is, not the result of chance alone. Statistical tests can help you understand the meaning of the numbers of responses you received and show you patterns among the data. Statistics show you what is important about your findings and how you can use that information to interpret the results of your research.

Statistics are facts derived from your data. They summarize your findings into manageable figures, equations, or statements. Statistical methods help you organize and interpret what you've observed, recorded, or gathered. They provide a shorthand way of describing data and the relationships among individual pieces of information and of generalizing your findings to the group that you studied.

In this chapter, you'll develop a vocabulary of terms associated with statistics and statistical methods. If you'll regularly work with statistical information, you should take a course or further study the use of statistics. If you'll only infrequently work with statistical information, the vocabulary and basic understanding of statistical concepts and practices covered in this chapter should help you work with the statistical experts with whom you consult.

Descriptive and Inferential Statistics

Put simply, a *statistic* is a fact that explains or describes the results of observation. *Statistics* is the term used to describe the methods of making sense of quantitative, or countable, data. These methods are used to summarize, organize, evaluate, and interpret data. There are two broad types of statistics: descriptive and inferential.

Descriptive statistics organize and summarize data. They describe the characteristics of a study. For example, information in graphs often shows how data are distributed across a range of values. The graph may show the frequency with which certain data are distributed among values. Frequency distributions and illustrations of data are examples of descriptive statistics. When you figure an average, a mean, a median, and the standard deviation of values, you're working with descriptive statistics. They help you organize data into meaningful statements or descriptions and present your results in a condensed format. Descriptive statistics may be used to describe a single variable or to compare variables.

Inferential statistics are used to evaluate and interpret data. Often you're trying to determine a cause or an effect. When you work with inferential statistics, you evaluate what you've learned about a group (the sample) and then generalize what you've learned to the larger population from which you took the sample. Some statistical tests that help you make these generalizations about a larger group are the *t*-test, chi-square test, and the analysis of variance (ANOVA). Inferential statistics help you draw conclusions from your findings.

Variables

Statistics describe the relationship of variables in a study. (You may have read a brief description of variables in Chapter 11, Empirical Experimentation.) When you conduct an experiment, you determine what you want to study, which variables will be studied, and how they have an impact on other variables. When you later conduct statistical tests, the number and type of tests you choose depends on the type and relationship of variables in your study.

You might have worked with independent, moderator, and dependent variables. In your study, you may have manipulated an independent variable to study the effects of that variable on other, dependent, variables. When you conduct a formal experiment, you analyze how an independent variable influences a dependent variable. Perhaps you conducted a usability test, for example, to see how many function keys computer operators remembered when they were asked to complete a typical data-entry task. You wanted to test the user's recall of which keys were used in the task. Recall was the dependent variable in your study. You then wanted to test the relationship of a distraction on the user's ability to recall the keystrokes. The type of distraction is a manipulable variable; it's an independent variable. You might also have varied the conditions under which the test was given. Different versions of the test may have been used to test varying degrees of difficulty, in a cool or a hot room, or under a time limit. Time, degree of difficulty, and temperature could have been the independent variables that you might use to see their impact on the computer operators' recall of keystrokes.

Moderator variables provide a context for the dependent and independent variables. Moderator variables represent existing characteristics of the participants or materials used in the study. They can be used to classify participants or materials, and they're not manipulated by you. Gender, age, and amount of previous training could be moderator variables in the example.

Terms Associated with Descriptive Statistics

Descriptive statistics summarize data so that you can understand individual facts and the relationship among facts that come from your research. To describe the *central tendency*, or the typical characteristics of the data set, you can figure the mode, median, and mean.

If you gave a group of trainees a posttest to learn how much they remembered from the training session, you might have these raw data. Eleven trainees took the posttest, and the scores were 93, 87, 87, 87, 87, 80, 79, 76, 75, 73, 70.

The *mode* is the most frequent score or value within the data set. For example, you found from the results of the posttest that four trainees had a score of 87. Eighty-seven is the mode, the most frequent score among the data set (the number of posttests) from this sample (the group of trainees who took the test).

The *median* is the point above which half the scores fall and below which half the scores fall. If half the trainees (five) scored above 80 points on the posttest (with scores of 93, 87, 87, 87, and 87) and half (five) scored below 80 points on the posttest (with scores of 79, 76, 75, 73, and 70), the median would be 80. The median is often abbreviated as *Mdn* in statistical notation.

The *mean* is the average score in the data set. If the trainees' scores are 93, 87, 87, 87, 87, 80, 79, 76, 75, 73, and 70, the total of all scores is 894. Divide that number by 11 (the number of trainees), and the average is 81.27. The mean score is then 81.27, or however many decimal places it's appropriate to have to describe your findings. A mean is often abbreviated as *M*.

When you work with descriptive statistics, you also might be interested in the way figures are dispersed across a range of score. A *range* may be defined as the difference between the highest score and the lowest score, or you may also see it defined as the difference between the highest score and the lowest score, plus 1. In either case, the range describes the top and bottom of the scale across which your scores are dispersed.

You might also be interested in the *frequency distribution* of data in the range. Frequency distributions show where data are grouped within the range and if patterns exist among data. When you analyze a range, you're also interested in the variability of scores in the range. You want to see how much variation occurs among the data. The most common measure of variability is the *standard deviation*. In statistical notation, it is abbreviated as *SD*. The standard deviation describes the average distance (the deviation) from the mean score.

Another term used in descriptive statistics is *correlation*. A correlation shows what happens to one variable when another variable changes. Correlations can be positive or negative to show the direction of the change.

Terms Associated with Inferential Statistics

Instead of just describing the data you collected, you may want to infer, or generalize, a behavior or a type of response to the whole group you studied or the larger population of which the group is a part. Then you'll work with inferential statistics to evaluate the data and generalize your findings to a larger sample. You might, for example, want to see how a larger number of trainees might score on the posttest. You might want to determine the probability of other trainees scoring within a certain range. Many tests associated with inferential statistics relate to probability, or the predicted chance that something will occur, based on your analyses of what occurred with the sample in your current research project.

Some terms associated with inferential statistics describe the components of the tests of the data. You need to indicate how many items or people made up the sample. If you used subgroups within the sample, you need to indicate the size of those subgroups. After you've conducted statistical tests, you need to report how often the results can be attributed purely to chance, or how many of your findings are significant. A few abbreviations commonly listed in results of analytical tests used with inferential statistics are these:

df — Degrees of freedom, or the amount of variation expected among scores.

F-ratio — A measure of the within-group and between-group variance; a ratio used in analyses such as the analysis of variance (ANOVA) computed by dividing the variance between treatments (or experiments) by the variance within treatments.

N — The total number in your sample (for example, $N = 150$ means that you had 150 people or items in your sample).

n — The number in a subset of your sample (for example, $n = 15$ means that you had 15 people in a subgroup of the total sample of 150, as in the previous example).

p — A measure of probability that indicates how much variance among scores is due to chance only ($p \geq .05$ indicates that the probability of something occurring purely by chance is greater than or equal to 5 times out of 100).

t — A statistic figured for two samples to illustrate the difference in the sample's central tendencies, or typical characteristics.

Significance

An important consideration when you analyze findings is which findings are significant. In statistics, *significance* means more than important or noteworthy. Statistical significance is the unlikeliness that the interrelationships within the sample can be attributed to sampling error. When you conduct tests of statistical significance, you analyze the data to learn if your methods unduly influenced or skewed your results.

Depending upon the type of reserach you conducted, you can analyze the data using several different statistical tests. Tests of significance determine the probability that a relationship between two variables in the sample can be inferred to a relationship between the same two variables in the larger population from which the sample was taken. In other words, tests of significance help you determine what you can logically generalize from your results to the larger population.

P-values, or probability values, indicate the probability that the relationship between variables occurred by chance. An equation of $p \geq .05$ indicates that the probability of the relationship occurring because of chance alone is 5 percent, or 5 times out of 100. The degree of acceptable probability that findings were due to chance varies with the number of participants in the study and the experimental method. The determination of probability and what are acceptable probability levels determine when a finding is statistically significant; that is, it most likely didn't occur just by chance.

Statistical Tests

You conduct statistical tests to analyze the relationships among variables, evaluate the data, and determine the validity of your results. Depending upon the type of research and data, you might work with one or several statistical tests. Table 13.1 provides a brief overview of some tests you might use.

Computerized statistical analyses are available, and you can use programs or software applications designed to do the mathematical computations for you. Computerized analyses can save you a great deal of time, but you need to know how to interpret the results correctly.

Unless you're familiar with statistics, you should consult with statistical experts to determine the appropriate tests of your data. These experts can also help you understand the relationship among data and what they indicate about your results.

Table 13.1 Definitions of Common Statistical Tests

multivariate analysis	an examination of the simultaneous relationship among several variables (e.g., analyzing the relationships among age, income level, educational level, and gender)
path analysis	a type of multivariate analysis, in which the causal relationships among variables are represented in a graph
regression analysis	an examination of the relationships among variables and illustrating these relationships in an equation
univariate analysis	examination of one variable in order to describe it
bivariate analysis	examination of two variables simultaneously to determine the relationship between them
ANOVA	an analysis of variance to compare two or more treatments or two or more populations to learn if there are any mean differences among them
MANOVA	multiple analyses of variance to compare two or more treatments or two or more populations to learn if there are any mean differences among them
Chi-square	a goodness-of-fit test to determine how well the data fit the null hypothesis
one-tailed *t*-test	a test to determine if one group (the control group) had a numerically larger score than that of the other (experimental) group, or if the experimental group had a numerically larger score than that of the control group, according to the researcher's hypothesis
two-tailed *t*-test	a test to determine in which direction differences between the control group and the experimental group occurred, according to the researcher's hypothesis

Using Statistics in Documents

When you write the results of your research for publication, you'll have special guidelines to follow on the use of statistics. Some abbreviations noted previously in this chapter will be used in the equations and statements describing the kinds of tests that you ran on the data and the meaning of your results. Most style manuals include sections describing how statistics should be reported. For example, the *Publication Manual of the American Psychological Association* describes the appropriate abbreviations and the correct citation style for using statistics in scientific publications.

To follow APA style in your report of your research, you include information about the magnitude or value of the statistical test and note items such as *t*-tests, *F*-tests, *p* (probability) values, and degrees of freedom when you're discussing inferential statistics. When you include descriptive statistics, you should note means, standard deviations, and variances. Although some statistics can be reported in the text of the results section of journal reports, readers seldom enjoy reading line after line of statistical information. You should cite statistics in the text to indicate how your research was conducted and what results you obtained. If you need to provide a great deal of statistical information about the data gathered during your research, you'll probably need to include one or several tables to provide further statistical detail.

Table 13.2 Definitions of Terms Used with Statistics and Statistical Methods

Term	Definition
continuous variable	a variable that can have an infinite number of values between one variable and the next
correlation	the relationship between two variables
descriptive statistics	summarize and arrange data into easy-to-understand information
discrete variable	a variable with a finite number of values between one variable and the next
distribution	a set of scores
inferential statistics	statistics and methods used to make generalizations from the data about the sample population
nominal scale	a measurement scale in which data are labeled and categorized
ordinal scale	a measurement scale in which data are ranked by size or magnitude
population	every member of the group you want to study
raw score	an original measurement or an original value within the distribution/set of scores
sample	part of the group that you study
statistic	a single characteristic of the sample/group being studied
variable	what can change and have different values

Summary

Understanding statistics takes more practice and study than you can read about in a single chapter. However, you can develop a vocabulary of the terms and tests associated with inferential and descriptive statistics so that you can talk with statistical experts or begin to build a foundation on which further study into statistics can be based. As an overview of common terms, Table 13.2 provides a summary of common terms used in statistics.

Research Activities

1. Talk with statisticians and statistical consultants in your university, company, or research institute. Discuss the types of analyses that are appropriate for your current project.
2. Have a consultant conduct statistical tests, as appropriate, on the data from your recent research. Discuss the terminology and the meaning of statistics as they apply to your work.
3. Read the sections explaining how statistics should be reported in several style manuals. How are inferential statistics reported? How are descriptive statistics reported? When should you include statistics in tables? When should statistics be included in the text?

For Further Reading

You might read several textbooks dealing with the use of statistics in your technical or scientific specialization. A few textbooks are cited in the following list, but several other good books about statistics are available.

Babbie, Earl. 1995. *The Practice of Social Research.* 7th ed. Belmont, CA: Wadsworth.

Hogan, J. B. 1983. "Statistical Doublespeak: The Deceptive Language of Numbers." *The Technical Writing Teacher* (Winter/Spring): 126-29.

Krauhs, Jane M. 1993. "Extend Your Confidence Limits in Writing about Statistics." *Technical Communication* (November): 742-43.

Spyridakis, Jan H., Michael J. Wenger, and Sarah H. Andrew. 1991. "The Technical Communicator's Guide to Understanding Statistics and Research Design." *Journal of Technical Writing and Communication* 21(3): 207-19.

CHAPTER FOURTEEN

Usability Testing and Validation

Technical communicators are always concerned with recognizing the needs of the audiences to whom they are presenting information so that it is accessible to, usable by, and useful to that audience. This concept, so easily stated in such a few words, is not always so easy to practice. Audiences need and use information differently in different situations and different environments. Thus, what is useful content, or format, or style, or medium of presentation for one audience will not necessarily be useful for another, nor perhaps for the same audience working with the same material in a different environment.

These concerns are not new to technical communicators; thus, there have historically been various processes for conducting research to determine the effectiveness of a document both before the documents were released and while the audiences were using them. In industry, documents might be reviewed by subject matter experts (SMEs), who help ensure accurate content, and by editors, who help ensure clarity of style and presentation, as well as adherence to corporate and other standards. In some cases, procedures documented in instructions are tested by being used. Final, disseminated documents might have a form of usability survey or a return postcard for audiences to respond to the usefulness of documents. Many times, audiences can call 800-numbers to discuss problems with the usability of documents.

Each of these procedures is part of a continuing research effort designed to determine if a document is truly usable by the audience. Usability testing has emerged in the last few years as a major approach to extending the range and focus of methods we use to test documents in technical communication. In this chapter, you'll read about strategies for developing usability tests, including elements of those tests such as readability and validation.

Usability, Readability, and Validation

Usability in technical communication refers not just to paper documents, but to every type of scientific or technical information that is created. Online Help menus, wall posters, an interactive CD-ROM set of conference proceedings, a three-dimensional mechanical drawing, a keyboard template, a robotic model of the universe, a danger sign on a shopfloor, a journal article—each one is usable in a different way and must meet the audience's expectations for that kind of information.

To be usable, information must by definition be accurate and complete. It must provide audiences with as much information as they need and be correct. For example, a fourth grade student may need only definitions about the parts of a flowering plant and brief descriptions of the plant's life cycle. A botanist may need more in-depth information about different plants' reproductive cycles and the ways hybrid plants can be developed to create greater yields. A county agriculture agent may need to know how well hybrid plants will grow in different soil types and how plants react to specific chemicals. The information for each person should be complete for his or her needs, and the information should be accurate, not partially true or true up to a point.

Usable information also must be up to date. The most recent accurate information must be used. In technical communication, that often means that gathering information up to the time of publication or distribution is necessary, because some subjects change so rapidly.

Usable information can't be misleading. As many different accurate perspectives as possible must be represented through the information gathering, so that the audience gets a balanced picture of a subject. Information can sometimes be misleading because the person who created it didn't edit carefully or didn't carefully consider where images or words were placed. For example, a cutaway view of an engine may be an attractive graphic, and it may accurately illustrate the external and internal components visible when the engine is shown sliced in half. But technicians who need to see the relationship of one specific component to the rest of the engine might find an exploded diagram more useful, because the components are seen as exploded outward but can be visually "scrunched" together to show where each component fits in the whole engine. Or, take the following examples of word placement:

The engine may crack if you pour in water while it is running.
If you pour in water while the engine is running, it may crack.

The misplaced modifier in the first sentence creates a very different image of what will occur. In the first sentence, the water is running, which creates a convoluted image of what is being described. The following sentence, by its ambiguity, creates a misleading impression:

Do not add too much water to the coolant.

Or what? Readers don't know if something bad will happen if "too much" water is added, or if the coolant will just be very weak. And how much is "too much"? Finally, "Do not" is a negative phrase. Is there a positive way to tell readers what to do? For many reasons, this sentence is not usable. It's misleading, although the writer may have had a process clearly in mind when the sentence was written.

At other times, writers may deliberately mislead by choosing how information is portrayed. A minor loss during the past financial quarter may seem much more serious if the amount is colored bright red on a graph. The often heard statement that "anything can be proven with statistics" unfairly condemns a very useful method of explaining relationships among data. But the technical communicator who uses statistics must be careful to select the right tests to evaluate data and must carefully interpret the statistical information.

Usable information presents as true a picture as possible. In some situations, the information must be reliable, so it can be used over and over to achieve the same or a similar outcome. In other situations, the picture may be changing so rapidly that it is only valid one time. But at the time the information is presented, it presents a true picture.

Usability, then, can be defined as consisting of two components:

- Validation
- Readability

Validation is the test for accuracy; it is a verification of information. When information is valid, readers know if what they're reading is a fact (a proven statement), an opinion (at least one person's thoughts or perceptions), a hypothesis (a statement developed to test an idea), a quotation (someone's exact words), a synthesis (a summary of relevant information), and so on. Through validation, the information is tested to make sure it's correct (for example, the product works as the writer describes it, all the steps in the instructions are in order and lead to the desired outcome, the page numbers listed in the manual's table of contents match the pages on which the headings are found in the text).

Readability is much trickier to describe. It is the test for clear, precise, correct, manageable chunks of information. Readability thus includes tests for grammatically correct language, the effective use of color in graphics, the selection of a large enough font for easy reading, the layout of pages so information is appropriately highlighted, the technically correct shades of meaning in word choice, and hundreds of other aspects of language, design, style, graphics, and format.

Preferences and Empowerment

The physical characteristics of information and the presentation of information can be tested and quantified to see if they match the audience's needs and expectations. However, audiences respond to more than quantifiable information. They form opinions and attitudes about information. That is, they have preferences about the way information is presented, even when the information is accurate and complete. For example, some readers may prefer reports on recycled paper, although the information is equally useful and correct when it's printed on new paper.

Information should be inclusionary, not exclusionary. *Inclusionary information*, by its language, format, and style, attracts the appropriate audiences and makes them feel part of a group. It draws people together by its language and appearance. The language, for example, is not sexist, and so it makes everyone who reads it feel good about the language, whatever the content. Inclusionary devices could include the use of Braille documents in a meeting so that everyone can read the information at the same time during the meeting.

In contrast, *exclusionary information* deliberately excludes some people from the targeted audience. Using jargon that not all members of a group understand is an exclusionary device; those who understand the jargon become part of an "in" group, whereas those who don't understand it become a group of "outsiders." Of course, some information should not be presented to everyone; it's appropriate for information requiring a security clearance, for instance, to be coded so that people without the required clearance are excluded from seeing the document. However, more often than not, audiences feel excluded not because of the deliberate use of an exclusionary device, but because the writer failed to consider every member of the audience's needs.

The term *empowerment* is more frequently used to suggest opportunity and ability. People are empowered when they have equal opportunity to exhibit their abilities, to reach their personal goals and objectives. When you conduct a usability test, you want to consider how information that is honest, accurate, complete, and accessible is empowering.

Empowering information may be information that is phrased positively. *Do not* limits activity and inhibits audiences. The negative does not empower people, whereas *do* is an empowering verb. It helps people act and promotes constructive behavior. It allows people the opportunity to exhibit their abilities and reach a personal goal or objective. Therefore, instead of writing, "Do not park here," a better sign might indicate where to park. Part of usability testing, then, should include an understanding of how the information can be used to empower the audience.

Testable Elements of Hardcopy or Softcopy Information

When technical communicators test their information, they can measure the usability of several elements that make up the effectiveness of the information:

- design
- language
- graphics
- interfaces
- availability of information
- accessibility of specific information
- accuracy of information

You can test several different design elements in hardcopy and softcopy information (see Tables 14.1 and 14.2).

Table 14.1 Design Elements of Hardcopy Information

page layout	use of color	choice of font	choice of typeface
placement of graphics and text	type of paper	use of tabs	type of binding
use of italics	use of boldface	use and style of borders	use of indentation
use and style of borders	use and style of bullets	levels and style of headings	style and placement of captions
type of numbering system	type of lettering system	use and style of hot spots	use of appended materials
documentation to accompany other information	style of documentation based on a style manual (for example, *Publication Manual of the American Psychological Association*, *Chicago Manual of Style*)	number of graphics in relation to amount of text	number and style of navigational tools (for example, indexes, tables of contents, lists of illustrations)

Table 14.2 Design Elements of Softcopy Information

screen layout	use of color	choice of font	choice of typeface
placement of graphics and text	number and type of icons	use of navigational tools	number and types of links between windows and screens
use of italics	use of boldface	use and style of borders	use of indentation
use and style of borders	use and style of bullets	levels and style of headings	style and placement of captions and labels
type of numbering system	type of lettering system	use and style of hot spots	number of graphics in relation to amount of text
use and style of animation	use and style of video	use and style of clip art	use and style of sound (for example, music, voices)
use and style of music	use and style of sound effects (for example, beeps, animal noises)	loudness	amount of information on the screen at one time
number of types of information (for example, graphics, text, sound)	number of types of interaction with the information (for example, touch, voice recognition, keyboard, mouse)	complementarity of all windows on one screen at one time	amount and style of documentation to accompany other information

Table 14.3 Linguistic Elements for Usability Testing

Linguistic Elements			
word choice	grammatical correctness	length of sentences	length of paragraphs
appropriateness of language (for example, nonsexist)	appropriate vocabulary	choice and use of organizational patterns	type of structure (for example, paragraphs, lists, hazard information)
sequence of words	use of punctuation	consistency of terminology	precision in meaning
conciseness	clarity	use of humor	authorial style

Language is another part of hardcopy or softcopy information that you can test, and again, there are many different elements to consider when you decide to test the effectiveness of language (see Table 14.3).

Readability tests can assess the word choice, average sentence length, document length, and educational grade's reading level needed to understand the information. Technical communicators can use standardized readability tests, such as Gunning's Fog Index, to learn how well their audience must be able to read in order to understand the information. Several different types of readability tests are available online, either as separate software or within word processing software, or they can be conducted manually.

Online grammar checkers can highlight long sentences, incomplete sentences, incorrect punctuation, passive voice constructions, and other common grammatical problems that may make information less readable. Of course, experienced technical communicators can use online checking programs routinely to provide a quick evaluation of their prose, but they should be able to distinguish when a grammatical rule should be followed exactly or when it is stylistically acceptable to bend or break the rule. Novice technical communicators may want to use online checking programs more carefully; they may want to read the explanations why something should be changed, but then make their own decision about the advisability of making that change.

In addition to checking the prose, technical communicators also need to evaluate the visual and aural effectiveness of their information. Graphics

Table 14.4 Graphic Elements for Usability Testing

Graphic Elements			
type of graphics	number and style of graphics	captions	documentation of graphics
use of color	use of text	use of sound	degree of movement
degree of interactivity	consistency in style among graphics	callouts and leader lines	appropriateness of perspective or view
pace of movement	sequencing of images	size of graphics	choice of font and typeface in text

consist of tables and figures in hard copy, but in soft copy they may be either static (nonmoving tables or figures) or moving. Users also may be able to interact with softcopy graphics, for example, to rotate a mechanical drawing, stop or start animation, or change the data provided in a chart. Some graphical elements that can be tested are found in Table 14.4.

When technical communicators plan tests of design, linguistic, and graphic elements, they often check what has been created, where it is placed, and how consistently it's used. They match the audience's needs and expectations against the technical communicators' original research about what the audience needed and wanted. And if the style, structure, and format don't match the audience's real needs and expectations, these technical communicators revise the information until it does match.

As you read these lists, you can probably think of more elements. When usability tests are conducted, a whole test can be developed to check the effectiveness of one or several elements. Although hardcopy and softcopy information are being emphasized in this chapter, you should remember that technical communicators may be working with many different media, including film, video, and presentations. These formats may involve additional elements to test.

But usability involves more than just making sure the information has been created properly. It requires that technical communicators ensure that people will be able to use the information, and to interact with it in appropriate ways. It also involves making sure that users can find the information they need when they need it and where they expect it. And it ultimately involves still more checking to ensure that the information is accurate.

To test the effectiveness of interfaces, technical communicators may need to evaluate the number and types of interfaces; the navigational tools, such as icons, directional arrows, and prompts; and the placement of these navigational tools.

But just creating and designing various forms of information are not enough. Information must also be available when users need it and accessible in ways they can easily, or intuitively, understand. To test the availability of information, technical communicators may decide to test how much information is needed by a particular audience at a particular time. For example, they may need to test whether users can find Help information when they have a question about a task they want to perform with their computer program. Then they also need to test the accessibility of that information. They may want to determine if the information is accessible in the proper format for that audience. For example, non-native speakers of a language may need information in their native language or may prefer less text and more graphics. Users who are working with online information need to have a good idea of where they can find information. They need to know, for instance, not only that Help information is available, but where and when they can access that information.

Validation, or the test for accuracy, might involve a separate series of tests or be part of a test for another element. For example, technical communicators might devise a test of a pie graph to see if users accurately interpret the information, or if the information is visually misleading. Technical communicators might test the language used in a set of instructions to see if users can interpret each step accurately, or if the word choice and organization are ambiguous or confusing. And, of course, validation also includes the correctness of the information. Technical communicators may devise performance tests to see if the information works as it's supposed to, if the information is in the correct order, is complete, and is up to date. In scholarly writing, for example, technical communicators may double check the accuracy of references to previously published materials.

Technical Communicators Who Conduct Usability Tests

Usability testing is often a team project, but who conducts tests, how often, and when during the information-development cycle the tests take place differ among companies and industries. But most technical communicators agree that usability testing in some form is a necessary part of the production cycle.

Who conducts usability tests? Depending upon the complexity of the information and its design, some information may be tested in several ways, involving several different members of the production team. Some companies prefer to have a separate department that conducts usability tests; other companies rely on the production team, with technical writers, technical editors, graphics specialists, production personnel, and technical experts to test the information before distribution or publication. The people who may be responsible for usability testing are

- the technical writer
- the technical communication and production team (technical writers, technical editors, graphics specialists, SMEs, design specialists, and production staff)
- an in-house usability testing group or department
- an outside usability testing consultant or company

Technical writers often conduct their own informal usability tests throughout the development of the information. They usually edit and revise their work several times to make the style more effective or to update the data. They sometimes use computerized readability and grammar checkers to help them revise their work. They may informally ask colleagues or other members of the technical writing group to try out the information, especially if they're writing procedures, instructions, or training materials. And they frequently conduct other validation tests by checking with SMEs or having SMEs review the information to make sure the content is correct and complete. These informal tests are useful during the development of the product, and they help ensure that the information is in good shape when it is reviewed by technical editors and perhaps further tested more formally later.

When formal usability tests are conducted, technical writers may be responsible for testing the information they've produced. They may be asked to develop usability tests, select the people who will test the information, and then use the results of the tests to improve the current product before it is distributed.

Because formal usability testing often involves a great deal of time and preparation before a valid test can be conducted, technical writers may not work alone. In many companies, a team made up of technical communicators, production staff, and technical experts develops and conducts the usability tests. Each expert brings his or her particular interests and areas of expertise to the test's development, so several different aspects of the information can be tested. Each team member may be responsible for developing questions or setting up sample tasks, for example. Then the team puts together the test based on each member's questions or examples. A team approach is a good way of testing several elements of the information with one test.

Usability tests developed and conducted by the person or people responsible for creating the product may be especially useful for testing one product, such as a manual, before the document is produced in final form and distributed to users. And at least at some level, every piece of technical communication should be tested before the audience receives it.

Ongoing usability testing, however, is equally important. An in-house usability testing group or department may be responsible for testing one or a few elements of all new information developed for outside distribution, instead of thoroughly testing all aspects of each product. An in-house usability testing group usually has expertise in developing reliable, valid tests that may be used several times. As each test is conducted, the group can learn more about what makes an effective product and how changes made to new information have improved the style, format, organization, and so on. Ongoing usability tests not only indicate the strengths and weaknesses of the current product, but they have an impact on the way information is designed in the future. For example, if a company changes the background color of its Help screens from blue to green and finds that users have difficulty reading green screens, this finding not only affects the current green-screened project, but future Help screens. When usability tests are conducted by the same department, whose only responsibility is to ensure that valid, reliable tests are developed, conducted, reviewed, and the results distributed, long-term trends, preferences, and improvements are more likely to be tracked.

Small companies may not have an in-house usability testing group, and when they need additional expertise in developing a test, they may call on a consultant or an independent agency that specializes in usability testing. Even large companies that only periodically produce information may prefer to hire an independent firm when it's time to conduct usability tests. These experts usually offer a variety of different tests and may administer them at their site, the company's site, or within a real work environment at the users' site.

Whether a single technical writer or a larger group is responsible for the tests, the appropriate people develop the usability test(s), conduct the test(s), and evaluate the results in light of what they mean to the current product and to the ongoing improvement of all products. But one other group is equally important in usability testing: the real users who may be asked to test the information. Usability tests must involve the audience that will eventually use the information. Although technical researchers, writers, and editors initially conduct audience analyses to determine what the audience expects, needs, and wants, usability tests ensure that these analyses are accurate and that the product is the result of these analyses. To determine the effectiveness of the information, you have to ask the people who work with that information—the users.

Conducting a Usability Test

As with any type of empirical research, you need a plan. If you're going to develop a reliable, valid test instrument, such as a questionnaire or a multiple choice test, you must follow a well-planned procedure. Conducting a usability test is no different from any other type of empirical research. You may have a hypothesis that, for example, placing a summary of key points in a blue box in the margin of each page will help technicians review important information and quickly skim the text. To test that hypothesis, then, you'd have to determine the best way to compare the way technicians read and remembered information from pages without the blue summary boxes with the way technicians read and remembered information with the blue summary boxes. Or you may need to test the effectiveness of online Help documentation, so you need to develop a way to see how quickly users can solve problems by accessing and following the documentation. A good usability test requires careful planning, if the test is going to provide you with useful data about your product's usability.

This procedure provides an overview of a plan for conducting a usability test. Of course, each task involves several specific subtasks and may require several people to help with the test (see Figure 14.1).

Plan the Responsibilities for Usability Testing

As with any research project, you must delegate the responsibilities for each part of the test. If you're the only person conducting the test, you still may need assistance, say, for the statistical analyses of the data or proctoring the test. Before you begin, you need to know who will do

Procedure for Conducting a Usability Test

1. Plan the responsibilities for usability testing.
2. Research what needs to be tested.
3. Create a user profile.
4. Set objectives for the usability test.
5. Design the test protocol.
6. Test the protocol on a sample group.
7. Select your test subjects.
8. Create the appropriate documentation for the test.
9. Establish the test environment.
10. Conduct the test.
11. Evaluate/analyze the information.

Figure 14.1 Conducting a usability test.

which tasks, where they'll complete them, and when they'll complete them. You might ask these questions as you plan each person's responsibilities:

- Who will be involved in creating the test? When? Where?
- Who will determine the testing criteria? When? Where? How?
- Who will conduct the test? When? Where?
- Who will analyze the test results? When? Where? How?
- Who will integrate the information into new designs? When? Where? How?

Each of these tasks can be broken into smaller tasks, too, but each question covers an important part of the testing process.

Research What Needs to Be Tested

If you could, you'd probably like to test every element of your information, just to see how your work fares with real users. Such a comprehensive test would be impractical for most projects, however. That's why you have to narrow the focus of your test to determine a hierarchy of important elements to test. Then you can select which item or items are the most important to evaluate at this time.

For example, the focus of a usability test might be one of the following:

- If users understand the technical terminology through in-text definitions.
- If users respond well to the music cues throughout the CD-ROM program.
- If users can complete a task by following the instructions.
- If users can locate information listed in the Index from within the document.
- How quickly users can locate specific types of information.
- How frequently users relied on Help documentation.
- How long users took to complete a tutorial module.
- How much readers remember after completing a tutorial.
- How much readers retain after reading a selection.
- How readers interpret information presented graphically.
- How well users liked (music, sound effects, visuals, color schemes, layout, and so on).

Any one of these items could be the emphasis of a usability test. If you want to test more than one item, you could evaluate similar items, such as how long users take to complete a tutorial and how much they remember after they complete it, with the same test.

You might want to find out which types of usability tests have been conducted on similar documents within your company to help you decide the focus of the current test. If your specifications for this project reflect a change in style, format, design, medium, organization, or some other element, you may want to test the effectiveness of that change. If several tests have been conducted about the same element, the usability test you develop may be a follow-up to previous research. Or you may want to test something completely different, because you have data about other elements that have been tested.

Because the type of test, the testing environment, the way the test is administered, the type of results, and other factors depend upon what you want to measure, you need to be very clear about your testing priorities before you proceed.

Create a User Profile

Before you select typical users to test the usability of the information, you need to review the audience analysis you completed when you first conducted research about what you would write or design. If the person or group conducting the usability test hasn't analyzed the audience (for example, an outside group may not know who your information users are), you may need to provide information about the target audience. When you (or someone else) creates a user profile, these questions are important:

- Who are the users?
- What do they need from the information?
- How familiar are they with the subject matter?
- When do they use the information?
- When do they need the information?
- What are their ages?
- What are their educational backgrounds?
- What are their cultural backgrounds?
- What are their preferences?
- What are their constraints?
- What are their expectations?
- What are their complaints about previous similar products?

When those who are developing the usability test clearly understand the people who will be using the product, they're more likely to create an effective test.

Set Objectives for the Usability Test

With all the preliminary planning out of the way, you can establish what you want to measure. For example, do you want

- To measure users' understanding of the terminology?
- To determine if the chunks of online information are arranged in the way(s) users think about and approach a task?
- To discover if users can find the information they want when they want it?
- To ask users the format they prefer?

Once you know what you want to test, you have to plan the test in relation to a measurable objective. You figure out how you can get the answers you want by knowing what exactly should be measured. Are you interested in attitudes, preferences, or other emotional responses? Do you want to measure the number of correct responses to questions about the subject? Do you want to measure the time taken to complete a task? The answers to these sample questions lead to very different objectives and, consequently, to very different types of usability tests.

Design the Test Protocol

You may need assistance in developing a reliable, valid test instrument, such as a questionnaire, or a test method, such as personal interviews. Several test protocols are available to you, but you have to know which type will help you meet your objective(s). Test protocols include

- interviews
- case study of on-the-job use
- job (performance) test
- description (aloud) of tasks as they are performed
- observation of subjects performing tasks
- written tests (such as quantitative measures, read-and-locate tests)
- online tutorials
- performance (timed) tests
- performance (accuracy) tests
- field surveys

Once you determine the type of test that best suits your needs, you develop the test and establish the methods for conducting the test. For example, you may know that you want to get users' impressions of a pocket reference card, so you decide that interviews with typical users will be most effective. You want real responses, and you want the users' own words. You decide to provide users with the pocket reference card at the start of the interview and then to record users' responses as they look at and use the card. Users must give their informed consent to participate in the research; they must know that you are going to record them and use their responses to evaluate the card. Before you let interviewers loose to talk with users, you'll probably want to conduct a training session so that the interviewers know they're not supposed to prompt responses, they avoid asking questions except for clarification, and they know how to gain informed consent. The protocol—the testing method and test itself—takes the most time to plan, develop, and refine. And after the next step in the procedure, you may need to go back to this step and revise the protocol again.

Test the Protocol on a Sample Group (Usually In House)

Before you select the test group and conduct the usability test, you should practice on a sample group. You might want a sample group of users, but more likely, you'll test the protocol in house. If you're using an in-house group, you may want to select employees who haven't worked on this project or who have very different job responsibilities from technical communication. You should try to find people who are as close as possible to the user profile you created earlier.

The practice test gives you the opportunity to work out any problems with the method or to refine the test itself. It gives you a better idea of the ways users will respond when they test your information's usability. Armed with feedback from the practice test, you may need to back up a step or two in this procedure until you've developed a better protocol.

Select the Test Group

You should follow the appropriate sampling procedures to select the participants for the test. Depending upon the protocol, you might want to select participants randomly, or you may want to select participants who work together at the same site. In general, the participants should be people who

- Can keep information about the product confidential until it is released or distributed.
- Match the profile of the target users.
- Are available to be tested (at a single site, on a given date).
- Are unfamiliar with this version of the product.
- Are objective about the test.

Create the Appropriate Documentation for the Test

After you've selected the participants and prepared the protocol, you should create any explanatory materials, such as legal releases or informed consent notices, that will be needed to document the test and provide instructions. Demographic measures and confidentiality agreements, for example, may be needed before you conduct the test.

Establish the Test Environment

The site should be prepared for the test. If you're simulating a work environment, you should have the appropriate equipment and arrangement of equipment and furniture, for example. If you're conducting the test in the users' work environment, check with managers at the site to ensure that the arrangements have been made for the test and that the space has been reserved for your use. You may need to visit the site, if you're not familiar with it. You should provide and check whatever is needed to create an appropriate testing atmosphere.

Conduct the Test

You have the participants, the site, the test, the method, and the personnel to conduct the test. Now, you just do it. You conduct the test, gather the raw data, and prepare them to be evaluated, following the procedures you established as part of the test protocol.

Evaluate/Analyze the Information

Depending upon the type of data you collected, you may need assistance in analyzing and interpreting the data. For example, statisticians can direct you to the appropriate types of statistical tests for the type of test you conducted and the data you collected. If you're evaluating the data, you have to make sure that you're analyzing them objectively and then determining how this information can best be used to improve the usability of your product.

Usability tests are empirical research, but they are not usually conducted for their own sake, just to gain more knowledge about the usability of certain elements in information design. They most often have a practical application, so that the current information can be revised before it's distributed or published. But it also has an ongoing impact on the way future projects are designed and how the information ultimately is presented to users.

Summary

Usability includes both readability and validation. Although the information must be accurate, it also must be in a format and style that are easy to use, organized so that specific pieces of information are easily found, and accessible to users when they want it. By conducting formal usability tests, technical communicators can check the usability of a single product (such as a manual) or develop ongoing procedures for evaluating and improving all information. Conducting an effective usability test means that you approach the test as an empirical research problem, and you develop the test protocol as carefully as you would develop any other type of research instrument and methodology.

Research Activities

1. Informally validate a current project by reading it "as a user" and testing the accuracy of your information. Then have a colleague informally validate the information and provide you with comments.
2. Plan a usability test for a current project. Involve several people who have worked on the project and/or who have special expertise (such as in data analysis) to create a team. Follow the procedure outlined in this chapter, and break down each step into specific tasks assigned to different team members.

For Further Reading

Appleton, Elaine L. 1993. "Put Usability to the Test." *Datamation* (July 15): 61-62.

Boiarsky, Carolyn. 1992. "A Teaching Tip. Using Usability Testing to Teach Reader Response." *Technical Communication* (February): 100-02.

Dumas, Joseph S., and Janice C. Redish. 1993. *A Practical Guide to Usability Testing*. Norwood, NJ: Ablex.

Ehrlich, Kate. 1994. "Getting the Whole Team into Usability Testing." *IEEE Software* (January): 89-91.

Grice, Roger A. 1994. "Focus on Usability." *Technical Communication* (August): 521-22.

___. 1993. "Usability and Hypermedia: Toward a Set of Usability Criteria and Measures." *Technical Communication* (August): 429-37.

Nielsen, Jakob. 1993. *Usability Engineering.* Boston: Academic Press.

Parker, Rachel. 1994. "Don't Test for Usability: Design for It." *InfoWorld* (July 18): 58.

Queipo, Larry. 1991. "Taking the Mysticism out of Usability Test Objectives." *Technical Communication* (April): 185-89.

Rubin, Jeffrey. 1994. *Handbook of Usability Testing: How to Plan, Design, and Conduct Effective Tests.* New York: John Wiley & Sons, Inc.

Simpson, Mark. 1991. "The Practice of Collaboration in Usability Test Design." *Technical Communication* (November): 526-31.

Skelton, T. M. 1992. "Testing the Usability of Usability Testing." *Technical Communication* (August): 343-59.

Tipton, Martha. 1993. "Usability Testing for Multimedia." *CD-ROM Professional* (July): 123-25.

PART FOUR

Possible Publication of Your Results

By now, you've conducted research and evaluated your results. What do you do with the information you've so carefully gathered? Research is meant to be shared with people who need or want the information, but you can share your findings in a number of ways. One good way is through formal publication: in a journal or magazine, as a book, as a chapter in a book, as a manual, as a corporate report. Another way is to present your findings at a conference. Whatever you decide is an appropriate way to publish your information, you should share your findings.

Submitting your work for publication requires additional research. You need to find the appropriate publisher and prepare your information so that it can be considered for publication. Throughout Part IV, you'll get ideas for conducting the necessary background research about publishers and publications, as well as for preparing abstracts, query letters, proposals, prospectuses, and bibliographies. These documents help you inquire about possible publication, submit information to be considered for publication, and develop your research into usable forms—for yourself and for others.

Part IV consists of these chapters:

- Chapter 15, The Query Letter, Proposal Letter, and Prospectus
- Chapter 16, Abstracts
- Chapter 17, Bibliographies and the Documentation of Sources
- Chapter 18, Conducting Additional Research

Chapters 15, 16, and 17 describe documents you write when you submit your work for publication or presentation. Then you learn about some specific documents—abstracts and bibliographies—you can use as a way of distributing information about your research or as a way of gathering more information. Abstracts and bibliographies can be created at the end of your research cycle, or they can be used to help you begin a new cycle of research. Chapter 18 ends the book with ideas about where you can go when you want to conduct additional research and how you can begin the process again, based on the research you've previously conducted.

CHAPTER FIFTEEN

The Query Letter, Proposal Letter, and Prospectus

The results of good research should be shared. When you've conducted research—or sometimes, when you've just started planning your research project or have begun conducting research about an important and/or popular subject—you can start thinking about where to share the results. If you decide that your information would best be shared in print with a wide audience, your decision will require you to enter into yet another form of research—finding appropriate venues for presentation.

Usually, there are several publishers for your work. For example, if you've conducted research within the broad discipline of sociology, you may publish your research to an audience of your colleagues who are also researchers. In that case, you select from among dozens of scholarly journals. You know that researchers have to keep up with others' research, and so scholarly journals may be an appropriate place of publication. Or you may decide that your work requires a more in-depth treatment, and the research won't suffer if it takes longer to make the findings available to your audience. Or you may discover that you have a good idea, and your research to date is useful, but you might be able to hone your idea more if you present it at a conference. Then you can get immediate feedback from your colleagues, discuss the research and its merits, and probably get ideas for the next wave of research. You might modify your presentation or clarify information about which your colleagues had questions.

After you limit your selection to a potential publisher, you should make contact with that publisher. Writing a query letter helps you learn more about the publisher and indicates to the publisher an important possible new source of information. Even if you submit your work without a formal query letter, your initial letter to the publisher in effect becomes your query letter or letter of transmittal. Many journal publishers and most book publishers require that you submit a query letter first, and then, if your information sounds like something the publisher might want, you're encouraged to submit a version of the proposed publication.

Another document you may need to submit, depending upon the type of publication, is a prospectus that outlines the sections of your proposed document. If you're considering publishing a book or, in some cases, an in-depth article, you probably will be asked to submit a prospectus. On the basis of your prospectus, then, the information is accepted or rejected, or you're asked to modify it.

Finding avenues of publication for your research is an important final phase of the research process, because one reason for conducting research is to share your findings with others. In this chapter, you'll read how to conduct background research about possible publication outlets and how to write a query letter and a prospectus.

Researching Possible Sources of Publication

Before you submit your work for publication, you should conduct some background research about potential publishers. You should consider the importance of each of these factors when you first start looking for a publisher:

- subject
- depth of the information
- timeliness
- distribution
- use of the information
- profitability
- reputation of the publisher

Subject

It may seem obvious that you know the subject of your research. However, although the subject is clear to you, the way you present it can vary greatly, depending upon the audience you want to learn about your findings. Is your research topic general or specific? Does it span several disciplines, or is it based in one? Is your information theoretical or application based? What is the most important emphasis for your subject right now?

Although multiple submissions of the same manuscript or prospectus to different publishers at the same time is generally discouraged and is often considered unethical, you may be encouraged to write about your findings in different ways for different audiences at different times. You might first write your findings for colleagues and experts who need detailed information about your methodology and results. Later, if your findings are well received and a more general audience might find the information interesting or useful, you might write about the implications of your research or summarize the significance of your findings. That's why it's so important to determine who first should learn about your work, and then later, to determine if you should eventually write other articles, presentations, and so forth, for other audiences. Once you clarify the emphasis of your subject matter and the specific part of the audience to whom you want to make the information available, you can narrow your search for a publisher.

Many journals specialize in a few topics or approaches to studying a subject, especially if the readers are highly experienced experts or specialists, whereas others publish a wide variety of articles for a more general audience. Book publishers, although they may cover hundreds of titles and subjects, usually have departments or individual presses that specialize in scientific, technical, educational, self-help, how-to, and other types of information.

To learn more about the subjects covered by a publisher, check guides like *Writer's Market*, look at the subject index and publisher index of *Books in Print*, and read recent issues of journals. When you've selected some possible publishers, write to them for author guidelines. These guidelines tell you the publisher's specifications and the types of publications that are preferred (see Figure 15.1).

Depth of the Information

How much information does your audience need? If you're writing for colleagues, researchers in your field, or other experts, you may provide

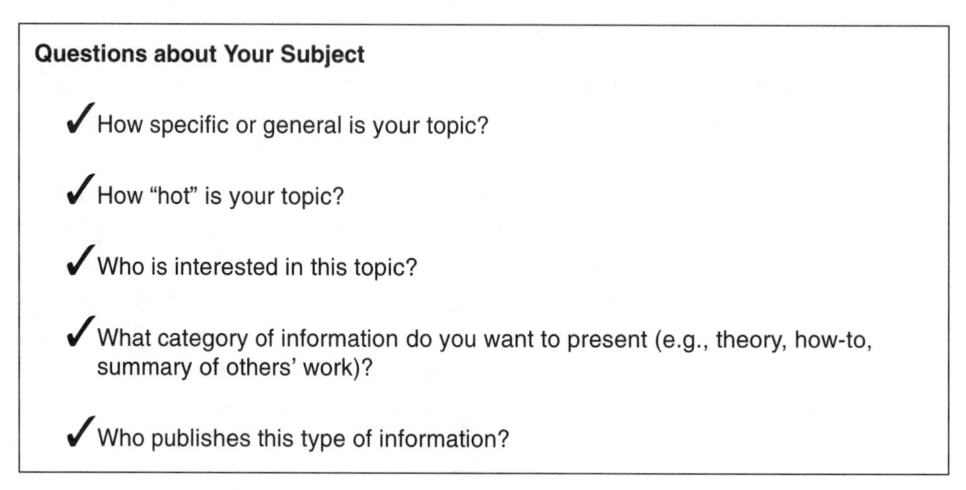

Questions about Your Subject

✓ How specific or general is your topic?

✓ How "hot" is your topic?

✓ Who is interested in this topic?

✓ What category of information do you want to present (e.g., theory, how-to, summary of others' work)?

✓ Who publishes this type of information?

Figure 15.1 Checklist for narrowing the subject.

a great deal of information, and it should be highly detailed. The audience may replicate your work, or at least they may want to evaluate your research methods, results, and interpretation of your findings. Books allow a longer treatment of a subject, and more supporting materials can be appended. If you need to provide an in-depth treatment of one experiment or project, a journal publisher may be a better option. Research or scholarly journals regularly publish reports or descriptions of recent research. If your subject is narrow and specialized, but your research was in depth, you probably are looking for a scholarly journal, perhaps in print form only. However, if your topic is about technology or involves state-of-the-art technical applications, you may find an electronic journal in which first to present information about your research. Then, later, you may follow up with a print article.

If your subject is application specific and confined to one application, you may also look for a journal or a magazine that publishes short, practical articles. Technical journals or the tips or how-to sections of popular journals are two places where you might submit a prospectus or a manuscript.

If your subject is application specific but deals with many applications, you may look for a book publisher. Some companies specialize in publishing how-to books, often as part of an ongoing series of publications.

If you have one important finding to share with others, but your work is still in progress and you plan to conduct additional research, you may look at the Calls for Papers for professional associations' conferences. The purpose of a conference is to share information about what is currently being done, what kinds of research are being conducted, and how practices and theories can be modified and improved. Conference presentations provide good opportunities to share information about work in progress, explain and describe what has already been done, and indicate plans for additional projects or research. Speakers usually receive feedback about their presentation, through written evaluations, questions and comments during question-and-answer sessions, and informal discussions after the presentation and throughout the conference. Sharing information formally, but often in a less threatening environment than that of widespread publication, is a good way to test your ideas, introduce your recent research, and get ideas for possible publications and for new projects.

Sometimes your work may involve several parts, and you may (ethically) be able to publish some parts of your information in different forms for different audiences who need different levels of information. In that case, you may need to decide how best to present your research in parts or in stages as it progresses, and how you might eventually publish your findings (see Figure 15.2).

Timeliness

How soon does your information need to be published? If your research topic is currently "hot," you may need a publisher with a short lag

Questions about the Depth of Your Information

- ✓ How broad or narrow is your approach?
- ✓ How detailed (or deep) is the information?
- ✓ How much depth will your audience need?
- ✓ Can the information better be published in separate parts or as a whole unit?
- ✓ Who publishes this type of information?

Figure 15.2 Checklist to determine the amount of information to share.

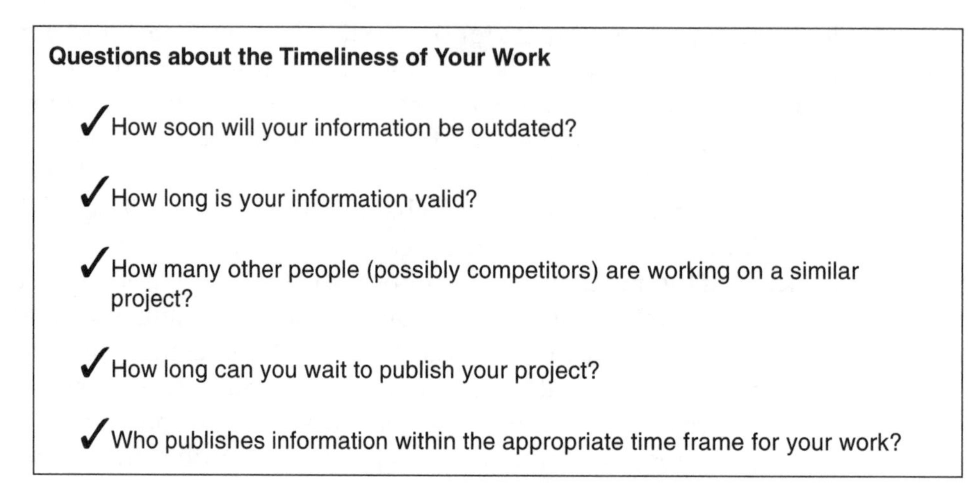

Figure 15.3 Checklist to determine a time frame for publication.

between acceptance and publication. Conferences, newsletters, in-house publications, electronic journals, and some printed journals have shorter publication schedules than other publishers. Book publishers and some scholarly journal publishers may have publication schedules of a year or longer. If your work is timely, you probably need to look at periodicals. However, if your work is relatively timeless and will be as useful in a year as it is today, you might consider journal and book publishers (see Figure 15.3).

Distribution

Of course, you want to distribute the results of your research to the widest possible appropriate audience. That's not the reason you conduct research, but it's the practical outcome of any good research project. How you distribute your information can have an important impact on the way the information is perceived and received.

If your work is distributed in paper form, it may be perceived as more traditional and long lasting. Paper information allows portability for readers, and it is often the preferred form of reference. Even with electronic information many readers still prefer having a hard copy of the information they deem most important or most often used. They like to have a paper reference to which they can return periodically.

If paper is the appropriate medium, you also must consider how readers will receive that paper. Journals published by professional associations often are sent only to their members and to the libraries that request the publication. People who are not part of that target group often may not know about or be able to find copies of the journal. However, if your audience is composed of people affiliated with a particular association, and these people are the experts who will find your work most interesting or helpful, then you probably don't need a wider audience.

Readers also perceive books differently from periodicals. Scholarly journals are generally considered more prestigious than popular magazines or shorter periodicals, such as newsletters, although each type of publication has its place in technical communication research. But books are still perceived as more important or more in depth than periodicals. Perhaps your work would make a better reference, one people can buy for themselves or can easily find in bookstores or libraries. Books tend to carry more weight, especially if they're produced by well-known publishers that specialize in your type of research.

No matter how popular or traditional paper documents are, some people now prefer to access electronic information. Electronic distribution of information is becoming increasingly important, especially because of its immediacy. Many more researchers and technical communicators have access to electronic information today than even a year ago. If your work should be distributed immediately to your peers, and in particular if your peers use electronic media daily, then you may want to consider simultaneous publication via networks and print.

Some professional societies and corporations publish two versions of their periodicals: one electronic and the other, paper. Or you may prefer to have your information distributed electronically so that almost anyone with an interest in the topic can learn more about your research, get in touch with you with their comments or questions, and provide the impetus for further research. Plus, if you distribute your information electronically, you may be perceived as more modern or technologically sophisticated, which are good attributes for technical communicators. Of course, as electronic publication becomes more common and more users have access to vehicles like the World Wide Web, the automatic prestige of electronic distribution will become passé.

You may also—or instead—want your information produced on CD-ROM and distributed via disk to subscribers of a journal or conference proceedings or to consumers who purchase disks commercially. If your work benefits from interactivity, and your research can best be presented in multimedia, then you may consider more interactive, multidimensional forms of distribution.

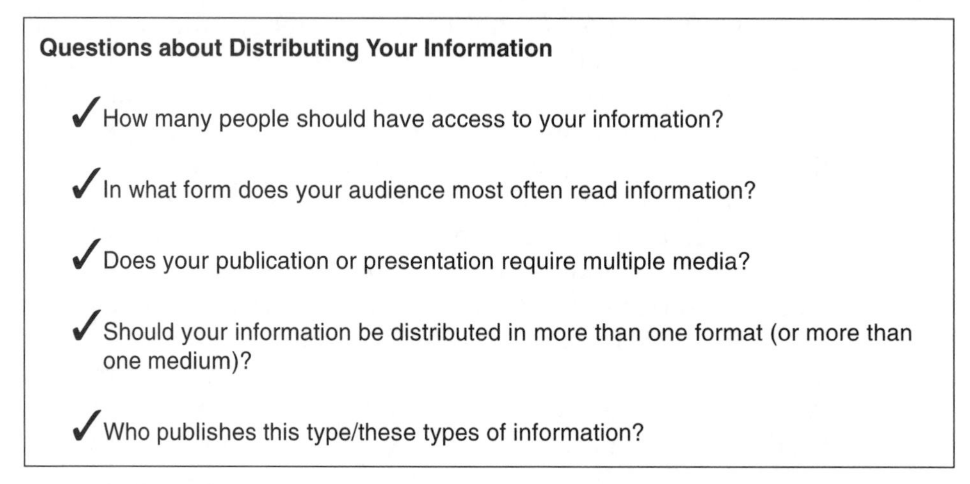

Questions about Distributing Your Information

- ✓ How many people should have access to your information?
- ✓ In what form does your audience most often read information?
- ✓ Does your publication or presentation require multiple media?
- ✓ Should your information be distributed in more than one format (or more than one medium)?
- ✓ Who publishes this type/these types of information?

Figure 15.4 Checklist to determine the distribution of your research.

Most readers have distinct preferences about the form in which information is distributed. They may like free access to information on a network, or they may prefer to pay for an electronic or a print subscription to a periodical. They may like their information in a linear, paper structure, one that can be used as a permanent reference anywhere. Before you submit information for publication, conduct an audience analysis of the people who need and will appreciate your information. Then you should know the distribution path your work should take (see Figure 15.4).

Use of the Information

A significant part of your research involves audience analysis, your consideration of how people will use your information. If your work leads to a practical application, such as training people to perform a task, helping managers make a decision at a meeting, and so on, then you have a good idea of the format your research should take when it's presented. You also know the methods of distribution that are best suited for the type of information, and the time frame in which such information must be made available.

But if your research is not so clear-cut, you may have to think more carefully about how people will use your information. Will they replicate your work? Will they use your work as a foundation on which to design their own research? Will they skim your information to keep up to date with what's going on in their profession? Will they discuss the informa-

Questions about the Way Your Information Will Be Used

✓ When will people need to use your information?

✓ Where will people need to use your information?

✓ How will people most commonly use your information?

Figure 15.5 Checklist to determine how your information will be used.

tion and possibly want to debate or refute it? Will they use the information as a standard reference? How people will use your information can tell you not only in what form ultimately your work will be expected, but also the marketability of your information (see Figure 15.5).

Profitability

The aim of your research may not, and in many cases should not, be to generate profit. However, some types of scientific or technical research result in a marketable product. Profitability, therefore, may be an additional consideration as you look for a publisher.

If your work is theoretical, if it is empirical and designed to promote additional research or expand the current body of knowledge, or if it is practical and can be applied generally for the public good, you probably will not charge a fee to anyone who wants your information. You simply publish it where the greatest number of appropriate readers can find it, such as in a scholarly journal published by a non-profit professional society. Or you may relay it to multiple users through a computer network, just to distribute information that will be useful to a number of other technical communicators.

But at times, your research is supposed to generate a product, and that product may be a potential profit maker, for you or your company. If your research is designed to create a new manual to accompany another product your company produces, the results of your research (which leads to the writing and production of the manual) are sold to make money for the company. You may have an in-house publisher, and so your search for a publisher is limited to your company.

At other times, you may market your research to an outside publisher, not only for the altruistic reason of sharing information with others, but

Questions about Your Project's Profitability

✓ Are the results of your research expected or required to generate a profitable product?

✓ Can your information be given to the public free?

Figure 15.6 Checklist to determine the potential profitability of your research.

with the expectation that people will pay for that information. Freelance technical communicators, for example, may make their living by writing articles for magazines, journals, newsletters, and other periodicals, or by writing or editing books about technical or scientific subjects. The product that results from the research is a marketable document (or other form of presentation), and the researcher/creator of that product has to determine who is best suited to market the final product.

If your research produces a product (such as a document or a presentation) with the potential for profit, you will need to conduct a market survey to determine who, how many, and where people are located who will potentially pay for your work. Then you narrow your search for the most likely publisher, who will also conduct a market analysis of the potential profitability of your project.

Some ideas are wonderful and should be shared with others, but the market is so limited that only a few publishers are willing to expend resources to publish the information. Other publishers recognize that even though a market is small and rather exclusive, their regular consumers need that information and rely on them to supply it (see Figure 15.6).

Reputation of the Publisher

Whether your publisher is in house or external, you will want to know the publisher's reputation. Some publishers, because they specialize in certain scientific or technical topics, have a well-developed reputation for dealing in that area. Readers may know what to expect from that publisher and rely on the publisher's judgment to present only high-quality work. If your research is accepted by that publisher, then your work has the publisher's good reputation with your audience already in its favor.

The publisher's reputation has little to do with the size of the business or type of publications (electronic or paper) that it produces. It has to do

Questions about Your Potential Publisher's Reputation

✓ What kinds of publications does this publisher normally produce?

✓ How well does the publisher work with authors and others involved in the publication process?

✓ What is the quality of previous publications?

✓ How well known is this publisher among your readers?

Figure 15.7 Checklist for evaluating the publisher's reputation.

with the quality of previous publications and the publisher's reputation for dealing fairly with authors, editors, and other people involved in the creation and distribution processes. You can learn a lot about a publisher by looking at several samples of publications. You might also talk with other authors who have worked with that publisher.

You can get a good feel for the way you can work with a potential publisher when you receive a response to your query letter, prospectus, or even informal personal communication with the publisher's representatives. If you receive a prompt reply, plus any additional information you may have requested about publication specifications, you can tell there's some interest in your research. The style and tone of the letter or the personal communication with the publisher or editor indicate the publisher's willingness to work with you and to publish your work. If you receive a form letter only, you may have a more difficult time selling your idea to this publisher. Sometimes you may have to wait several weeks or months to get a response, or until a deadline for submissions—for example, for conference papers—before you learn of any interest in your work. After the deadline, or after waiting the amount of time as indicated in the writer's guidelines, reference works like *Writer's Market*, or Information for Contributors to journals, you should follow up on your proposal (see Figure 15.7).

Where Should You Submit Your Work?

In short, follow these guidelines as you develop a strategy for submitting your work for publication or presentation:

- If you're conducting original research, or if you're replicating someone else's experiment, you may submit an article to a journal specializing in your field.
- If you're researching the history or trends of a product or subject, you may write a book or a long, descriptive article.
- If you're researching possible solutions to a problem in your company, you may be assigned to write or revise an in-house publication, in which case your publisher is the company.
- If you work for a company but are encouraged to share your information with a wider group of professionals, you may also submit your information to a professional society or conference board for additional publication.

Any research you do can and should be published. One of your responsibilities is to find the appropriate publisher for that information.

Beginning the Publication Process

When researchers submit their research findings or their ideas for possible research to a publisher, they write a query letter to potential publishers. A *query letter* can also be called a *proposal letter,* because in the letter, researchers propose their ideas, the way those ideas can be described in an article or book for publication, and the time frame for completing the research and submitting a draft for publication.

A *prospectus* is more developed than a proposal letter, although it is based on the same concept of proposing an idea or submitting a query about a publication; a prospectus includes a market survey that compares the proposed publication to similar publications already on the market. This type of proposal or query letter is frequently used when researchers want to submit book-length manuscripts for publication. In this chapter you'll see short query letters and a prospectus suitable for submission to publishers outside the company.

The Query Letter

Different publishers have different views about query letters. Some publishers won't accept them; others will, but only when accompanied by substantial parts of the proposed content material, such as a detailed outline. Other publishers insist upon a query letter before you submit material.

Part of your research about publishers involves learning what your potential publisher wants. You can do that by looking at such works as *Writer's Market, Writer's Digest,* or *Publisher's Weekly.*

Also, you can go straight to the publication. Somewhere in the introductory information in journals, perhaps in the information for contributors, there will probably be some statement about whether the publisher accepts query letters. Some publishers also specify the type of information they want in a query letter, such as if the research was refereed, if the research was conducted with the approval of a human subjects review board, if you've reported your information in other publications or are considering submitting your work to another publisher.

You also can make some preliminary judgments about the efficacy of sending a query letter simply by knowing the frequency of publication. Publishers whose publications come out weekly, for example, are less likely to want query letters than those publications that come out quarterly or are one-time publications (such as books). Those who publish only timely information usually don't accept query letters; it takes too much time to review them, and the information proposed in the query letters would be dated before it could be published. For book-length manuscripts, a query and, more likely, a proposal or a prospectus is required.

The form your inquiry takes may vary, but always think of your query or proposal letter as a sales vehicle. You don't want to ask an editor *if* there is interest in your material. You tell the editors *why* the material is essential for them to publish: you show that you know the publication's audience, the publisher's approach to the material, what the publisher has published in the past, and how the proposed material fits into the publication or with similar publications (such as within a series of other books the publisher already markets).

A Sample Query Letter

The query letter in Figure 15.8 outlines a proposed article for a scholarly journal. It is shorter than a proposal letter or a prospectus but still includes a description of the work in progress.

A Sample Proposal Letter

The query letter provides some initial background information about the proposed publication, but a proposal letter usually gives even more

1 Augusta Drive
Bowling Green, OH 43402

Mike Markel
IEEE Transactions on Professional Communication
Boise State University
Boise, ID

February 1, 1995

Dear Dr. Markel,

I have recently conducted research about the use of teleconferencing in technical communication courses offered by universities. During my research, I learned that many professionals have had a limited exposure to teleconferencing, and many have misconceptions about what teleconferencing is and how it might be used to enhance educational programs and course offerings. I'm convinced that teleconferencing as a link among universities and among universities and businesses will become an important aspect of higher education in the next 10 years.

I'd like to propose an article for *IEEE Transactions on Professional Communication* to share my information. I want first to describe a brief history of teleconferencing and bring readers up to date about the uses of teleconferencing in education and technical communication. I plan to include a glossary of key terms about teleconferencing, as well as illustrations of the practical technical aspects of linking universities and/or businesses. Next, I can explain why teleconferencing is so important, and how it might be used in readers' business or institution. Finally, I want to describe the benefits associated with teleconferencing and provide an overview of where the telecommunications industry is headed with teleconferencing.

I can send you a draft of the article, if you believe it is appropriate for your publication. Please let me know if you're interested in the article, and the form in which you would like to see it. I can be reached through e-mail: LPORTER@OPIE.BGSU.EDU, fax: 419-XXX-XXXX, phone: 419-XXX-XXXX, or at my home address, which is on the letterhead.

I'm looking forward to hearing from you.

Sincerely,

Lynnette R. Porter

Figure 15.8 A query letter to a journal editor.

details about the proposed work. It, in effect, has to sell the idea for your proposed publication to the publisher. In a proposal letter, you describe the work in persuasive terms, support your proposed plan with references or other documentation, and indicate a need for the information. The proposal letter shown in Figure 15.9 was used to describe this book.

This proposal letter includes a description of important information the publisher would like to know: the working title of the proposed book, the audience and purpose, a statement of the potential market, competitive works already published, and a plan for completing the book. The authors' credentials, the reason why they chose that publisher, and a description of enclosed materials were also provided. The proposal letter can also serve as a cover letter in a prospectus.

In addition to being a clear, persuasive overview of what you want to publish, a proposal letter should also highlight the important information and be as short as possible. Although the letter typically is more than one page, it should be fewer than four or five pages, even with an outline. Publishers are usually busy executives who receive many such proposals throughout the year. Your proposal must stand out as being professional, direct, informative, and persuasive. You should highlight important information by using headings, boldface or italics, or bullets, for example, as appropriate. In the sample letter, notice how information is organized, so that the first paragraph gets to the point but is still a cordial opening. The important information is separated not only in different paragraphs, but with appropriate headings. The attachments can be separated from the body of the letter, if the publisher wants to show the information to others in a meeting, but the letter form is still used to surround the attachments and give a context to them.

Before you write a proposal letter, organize your thoughts according to your answers to these questions:

- What, specifically, do you want to publish (for example, a textbook, a popular how-to book, a book of research findings, a formal report, a guide, a handbook, a manual, a tutorial)?
- Who will pay to use this information?
- In what form(s) should the information be published?
- How long will it take to prepare the information for publication?
- How big is the market?
- Who makes up the market?
- What will be your competition?

May 21, 1994

Terri Hudson
Acquisitions Editor
John Wiley & Sons
2827 Lucky John Drive
Park City, UT 84060

Dear Ms. Hudson,

We have been revising a book we developed for a research course required as part of Bowling Green State University's technical communication program. This book was initially created because there was nothing on the market to describe research strategies and methodologies technical communicators use on the job. Even though there were other books about research, none emphasized technical communication and covered the broad range of research skills needed to be effective as a technical communicator within a company or as a researcher/writer working in a lab or research institute. The methods vary widely, and no one book covers the range of research skills needed in the profession of technical communication.

Late last year we noticed an increasing interest in the state of research in technical communication. For example, the Fourth Quarter 1992 issue of *Technical Communication*, the journal produced by the Society for Technical Communication and widely read within the profession, was devoted to articles about research needs and the state of research. The emphasis was on an increasing need to conduct research within technical communication, not just within the scientific and technical disciplines in which technical communicators work. Students in technical communication courses need a solid foundation in research methodologies and should be encouraged to conduct research, but practitioners also need to do research and do it well, and currently that's not happening as much as it should.

Practitioners and students who want to learn more about research as it is needed in technical communication have only a few, widely diverse sources of information. Textbooks, such as Houp and Pearsall's *Reporting Technical Information*, include information about library sources, interviews, and questionnaires, for example, but that information is only a small part of the textbook. Articles about research can be found in *Technical Communication*, the *Journal of Business and Technical Communication*, and journals within technical and scientific specializations, but they are widely scattered. Even a comprehensive issue like that in *Technical Communication* referenced mostly conference proceedings, and the articles dealt with specific topics that need to be addressed. These sources are good, but they are not suited to be a guide to research within technical communication.

There are a number of good research books available, but they focus on one area of research. Books about statistics, for example, cover only one topic useful to

Figure 15.9 A proposal letter for a book.

technical communicators, and most readers don't want to use five or six different books to get the scope of information they need. Even style manuals describe how to cite research and may provide a cursory description of the need for ethics in research, but they are not designed to address other research issues. The closest book we've found to meeting the need for research in technical communication is *Conducting Research in Business Communication*, which was published by the Association for Business Communication in 1988. However, it also fails to address topics like documentation plans, sources of information, ethics in research, and the variety of qualitative and quantitative methods used within technical communication.

We picture the book as part of your technical communication series. We plan to include end-of-chapter summaries and activities, checklists for different research methods, and end-of-chapter bibliographies of recent research articles.

We've enclosed an outline to provide an overview of the book's content and style. As you can see from the outline, we've developed a thorough book. We've hit the variety of research possibilities for a number of reasons: 1) We want to be complete and up to date. 2) We know that the book can be used in classes, but the variety of approaches will make it attractive to practitioners, specifically those who are members of the Society for Technical Communication and work frequently with academic programs. 3) We hope as well that the variety will help current teachers of technical communication focus on creating research courses for their existing programs, if none currently exists.

We have each enclosed a vita that illustrates our credentials as researchers, writers, teachers, and technical communicators.

Please call us at 419-XXX-XXXX or 419-XXX-XXXX with your comments and suggestions. We're looking forward to working with you again.

Sincerely,

Bill Coggin　　　　Lynnette Porter

Figure 15.9 **(*Continued*)**

- Why will your publication sell?
- Why is your publication especially well suited for this publisher?

Then you determine how you can succinctly present the answers to these important questions, highlight the information so that someone skimming the letter can find the most important points, and sell your idea.

The Prospectus

A prospectus is usually submitted when you have an idea for a book-length manuscript. The prospectus then includes a description of the book (similar to what you wrote in the proposal letter) and sample chapters or sections or an outline of each chapter or section. Although you will have completed similar research about the publisher and the company's special needs, you also must do the following:

- Discover what other books exist on the subject about which you want to write. Read them to find out how they handle the subject; what their audience, subject, and purpose are; what kinds of examples they use; and what their intended audience is supposed to do with the information presented.
- Discover how recently the information has been updated. Just because a number of books are on the market doesn't mean that they all are successful or still useful. Find out if you have a new twist on the subject or new information that few, if any, books currently provide.
- Find out how many books about similar topics your proposed publisher currently has on the market. Look for series of similar books, such as a technical communication series, a research series, and so on.
- Determine which section or chapter best represents the total work in terms of audience, scope, format, and style. You want to show your publication's strengths, so be sure to submit effective sample chapters. First chapters usually are not the best to submit with a prospectus, because they often contain background or introductory material. They may not be as content specific as later chapters, or they may have too much prose and not enough examples or illustrations.
- Determine a realistic time frame in which you can finish the publication and present the information in final form to the publisher. This can often be a tricky part of the process, because you may have several other projects going on at the same time, and you need to allow time for researching, writing, revising, and editing, plus the time for reading and correcting proof pages before publication. The publication process takes quite some time in most situations, and you want to be able to create a high-quality publication that you, your readers, and your publisher are pleased with.

Again, your background research pays off when you begin to write. You know how you want to organize the information, and you're ready to put it in a clear, persuasive format. When you submit a prospectus, you generally include these materials:

- A cover letter: The cover letter introduces you and your proposed publication. It is a shortened version of the proposal letter, because the prospectus itself will add the important information about the book-length publication, the market, possible competitors, and so on. The cover letter is similar to other letters of transmittal you write for submitting documents. The letter transmits your work, in this case the prospectus and accompanying materials, to your audience, the publisher.
- The prospectus: This document should be able to stand alone—for example, as a handout in a meeting—but it should also be the focal point of the packet of materials you're submitting.
- Supporting documentation: A bibliography of similar books on the market, possible reviewers of your sample chapters, your resume or vita, or other information that backs up your prospectus can be attached as separate documents. Each one should be clearly identified so that it can stand on its own, but it also complements other information in the packet.
- Sample chapter(s) or section(s): Representative chapters in as final a form as possible should be included to give an idea of the breadth, depth, style, format, and audience of your work, as you see it.

Most of this information should be in print form at this stage, although you might also include a disk with your sample chapters or sections. You might also have communicated with the publisher through voice mail or e-mail, or you may have discussed the project in person or over the phone. But when you submit a prospectus, the formal, preferred form of the information at this time is still print.

A Sample Prospectus

The prospectus shown in Figure 15.10 is a version of the materials submitted for this book. As you can see if you compare this book's final table of contents with the proposed table of contents, changes were made between the acceptance of the prospectus and the publication of the book.

Then, additional research gathered after the prospectus was accepted created the need for further changes in audience, style, tone, and format. The sources cited as possible competitors in the market changed, too, because some sources were no longer as important in the marketplace; but more to the point, many other competitors entered that market.

The prospectus helps publishers see your vision for a work. The more clearly you can illustrate that vision, the better the publishers can see if they should share that vision or suggest another, more appropriate publisher.

July 5, 1994

Terri Hudson
Acquisitions Editor
John Wiley & Sons
2827 Lucky John Drive
Park City, UT 84060

Dear Ms. Hudson,

Bill Coggin and I have been revising a book we developed for a research course required as part of Bowling Green State University's technical communication program. This book was initially created because there was nothing on the market to describe research strategies and methodologies technical communicators use on the job. Even though there were other books about research, none emphasized technical communication and covered the broad range of research skills needed to be effective as a technical communicator within a company or as a researcher/writer working in a lab or research institute. The methods vary widely, and no one book covers the range of research skills needed in the profession of technical communication. We are updating the chapters we've previously written, including new information about electronic (most often computer-based) media and documentation to meet specifications, such as ISO 9000. In addition, we are adding chapters about statistics and scientific data-collection and analytical methods.

In late 1992 we noticed an increasing interest in the state of research in technical communication. For example, the Fourth Quarter 1992 issue of *Technical Communication* was devoted to articles about research needs and the state of research. The emphasis was on an increasing need to conduct research within technical communication, not just within the scientific and technical disciplines in which technical communicators work. Students in technical communication courses need a solid foundation in research methodologies and should be encouraged to

Figure 15.10 A book prospectus.

conduct research, but practitioners also need to do research and do it well, and currently that's not happening as much as it should.

Practitioners and students who want to learn more about research as it is needed in technical communication have only a few, widely diverse sources of information. Textbooks, such as Houp and Pearsall's *Reporting Technical Information*, include information about library sources, interviews, and questionnaires, for example, but that information is only a small part of the textbook. Articles about research can be found in *Technical Communication*, the *Journal of Business and Technical Communication*, and journals within technical and scientific specializations, but they are widely scattered. Even a comprehensive issue like that in *Technical Communication* referenced mostly conference proceedings, and the articles dealt with specific topics that need to be addressed. These sources are good, but they are not suited to be a guide to research within technical communication.

A number of good research books are available, but they focus on one area of research. Books about statistics, for example, cover only one topic useful to technical communicators, and most readers don't want to use five or six different books to get the scope of information they need. Even style manuals describe how to cite research and may provide a cursory description of the need for ethics in research, but they are not designed to address other research issues. The closest book we've found to meeting the need for research in technical communication is *Conducting Research in Business Communication*, which was published by the Association for Business Communication in 1988. However, it also fails to address topics like documentation plans, sources of information, ethics in research, and the variety of qualitative and quantitative methods used within technical communication.

As Frank Smith, Editor of *Technical Communication* noted in the research issue in December 1992, "the future of technical communication as a discipline or profession depends almost entirely upon the continuing and increasing conduct of research, both quantitative and qualitative, both basic and applied, both academic and commercial" (p. 521). The book we're proposing, *Research Strategies in Technical Communication*, could be used as a textbook in a research course, an additional textbook to a technical writing or a technical editing course, or a how-to book for technical communicators who never received formal training in research. In addition to being a valuable textbook that could be marketed not only to technical communication students, but to students in the sciences, technology, and the humanities, the book would be a valuable tool for professionals and should also be marketed to that audience.

The book provides an overview of quantitative and qualitative research and primary, secondary, and tertiary sources of information, online information search strategies, and individual chapters about the methods used to create an experimental design, a questionnaire, informative interviews with SMEs, and observations. Other chapters

(continues)

Figure 15.10 (*Continued*)

describe basic concepts used in statistical analyses of data, possible uses for the research, and ways to publish or disseminate findings.

We envision the format for each chapter as including the basic texts, with new information about computer-based research strategies and uses of computers in the research process; end-of-chapter summaries and activities; checklists for different research methods; and end-of-chapter bibliographies of recent research articles. Each chapter will have sample research documents or illustrations of the research process.

We've enclosed an outline and sample chapters to provide an overview of the book's content and style. Later chapters become more specific about each type of research and the appropriate strategies used within technical communication. We've also included our curriculum vitae to provide you an idea of our credentials and experience.

The chapters currently are available on Microsoft Word 5.1 for the Macintosh, but they can be translated into other software. The graphics will be made camera-ready and are not currently available on disk.

The attached bibliography lists some useful research books, but they either do not describe information about technical communication, are becoming outdated, or would require additional books to supplement the information about research in technical communication. Textbooks with a chapter about research have not been included, because most technical writing books deal at least superficially with research. These books indicate the current market and its limitations and illustrate where *Research Strategies in Technical Communication* could have a place in the market. If the title is revised to reflect a broader scope than technical communication, we would create an even larger market.

As you can see from the outline, we've developed a thorough book. We've hit the variety of research possibilities for a number of reasons: 1) We want to be complete and up to date. 2) We know that the book can be used in classes, but the variety of approaches will make it attractive to practitioners, specifically those who are members of the Society for Technical Communication and work frequently with academic programs. 3) We hope as well that the variety will help current teachers of technical communication focus on creating research courses for their existing programs, if none currently exists.

Some strategies we've included are simple, whereas others are complex, but that's the nature of the strategy, not our handling of the strategy. Although we don't believe because of its size and range the book will be attractive to lower-level undergraduate teachers, we do believe it will attract both upper-level undergraduate and graduate-level instructors. Because it is a comprehensive text, it most likely will

Figure 15.10 A book prospectus. (*Continued*)

be used as the only textbook for a course, but it could also be used to supplement other textbooks used in technical writing or editing courses. Finally, it could also be used by professionals outside academia who need to learn more about specific research skills or the uses electronic media are playing in research in technical communication.

Please call us (Bill Coggin at 419-XXX-XXXX or Lynnette Porter at 419-XXX-XXXX) with your comments and suggestions. We're looking forward to working with you.

Sincerely,

Lynnette Porter

Selected Research Books on the Market

Alkin, Marvin C., ed. *Encyclopedia of Educational Research.* New York: Macmillan, 1992.

Allen, Mary J., and Wendy M. Yen. *Introduction to Measurement Theory.* Monterey, CA: Brooks/Cole Publishing Company, 1979.

Anderson, Paul V., R. John Brockmann, and Carolyn R. Miller, eds. *New Essays in Technical and Scientific Communication: Research, Theory, Practice.* Farmingdale, NY: Baywood, 1983.

Babbie, Earl R. *Survey Research Methods.* Belmont, CA: Wadsworth, 1973.

Barlow, David H., Steven C. Hayes, and Rosemary O. Nelson. *The Scientist Practitioner: Research and Accountability in Clinical and Educational Settings.* New York: Pergamon Press, 1986.

Barzun, Jacques, and Henry F. Graff. *The Modern Researcher.* San Diego, CA: Harcourt Brace Jovanovich, 1977.

Campbell, Patty G., ed. *Conducting Research in Business Communication.* Urbana, IL: Association for Business Communication, 1988.

Hamburg, Morris. *Basic Statistics.* San Diego, CA: Harcourt Brace Jovanovich, 1985.

Hammersly, Martyn, and Paul Atkinson. *Ethnography: Principles and Practice.* London: Tavistock, 1983.

(continues)

Figure 15.10 (*Continued*)

Lauer Janice M., and J. William Asher. *Composition Research: Empirical Designs.* New York: Oxford University Press, 1988.

Marshall, Catherine, and Gretchen B. Rossman. *Designing Qualitative Research.* Newbury Park, CA: Sage, 1989.

McClelland, Ben W., and Timothy R. Donovan, eds. *Perspectives on Research and Scholarship in Composition.* New York: MLA, 1985.

Miller, Delbert C. *Handbook of Research Design and Social Measurement*, 5th ed. Newbury Park, CA: Sage, 1991.

Patton, Michael Quinn. *Qualitative Evaluation and Research Methods*, 2nd ed. Newbury Park, CA: Sage, 1990.

Tuckman, Bruce W. *Conducting Educational Research.* San Diego, CA: Harcourt Brace Jovanovich, 1988.

Yin, Rober K. *Case Study Research: Design and Methods.* Newbury Park, CA: Sage, 1984.

Outline of
Research Strategies for Technical Communicators

Part I. Introduction to Research Strategies in Technical Communication

1. Introduction to Research

Primary, Secondary, and Tertiary Sources of Information
Research Methodologies
Summary
Research Activities
For Further Reading

2. Idea Generation as You Begin Your Research

Writing and Researching in a Company Environment
Getting Ideas for Your Own Projects
Idea Generation with Computers
Summary
Research Activities
For Further Reading

Figure 15.10 A book prospectus. (*Continued*)

3. Documentation Plans in Corporate Environments

Information in the Documentation Plan
A Sample Doc Plan
Analysis of the Doc Plan
Documentation of Projects to Meet Specifications [such as ISO 9000, 9001, 9002]
Summary
Research Activities
For Further Reading

4. Ethics in Information-Gathering Activities

Ethics in Research Methodologies
Ethics Review Boards for Experimentation
Ethics in Interviews
Ethics in Questionnaires
Ethical Use of Secondary Sources
Ethical Review Process
Summary
Research Activities
For Further Reading

Ethics can be viewed in two different perspectives, as we address ethical concerns about different types of research. When technical communicators interview SMEs to gather information, they must work ethically by accurately recording what is said, using the information in context, and respecting their interviewees. When technical communicators create questionnaires, they must write questions that are as free from bias as possible but yet can elicit the type of information needed. When technical communicators use secondary sources, they must not plagiarize, misquote, or misinterpret the sources. These practices are part of ethical behavior in the day-to-day research needed for writing technical information. However, in addition, researchers who are designing experiments and conducting case study analyses must work ethically in these and different ways. They must protect their subjects' physical and mental well-being, and sometimes their privacy, as well. They often have to explain their proposed research methodologies to ethical review boards before approval for the research can be given. In this chapter, we discuss ethics from both perspectives and provide practical suggestions for ensuring ethical behavior.

Part II. Ways to Conduct Research

5. Computerized Information Retrieval

COM and Online Card Catalogs
Networked and CD-ROM Databases

(continues)

Figure 15.10 (*Continued*)

Online Search Strategies
LANs and WANs (Networks)
E-mail
Videoconferencing
Online Collaboration
Summary
Research Activities
For Further Reading

We describe computerized methods of locating and retrieving information, including the use of CD-ROM, networks, multimedia, and hypermedia. We also provide guidance for developing research strategies to use newer media effectively.

6. Personal Experience and Knowledge

Personal Experience
Personal Knowledge
Personal Observation
Summary
Research Activities
For Further Reading

We provide questions and examples so that researchers can evaluate their current knowledge and experience and develop strategies to gain new information. We discuss ways to observe SMEs on the job and methods of gaining experience with a product or a process so that technical communicators who will be writing about a subject area can do so confidently and accurately.

7. Experimentation

Ideas for Experiments
Literature Reviews
Hypotheses
Experimental Conditions
Experimental Protocols
Qualitative Methods
Quantitative Methods
Results
Summary
Research Activities
For Further Reading

In this chapter we define the scientific method of inquiry and the way to develop testable hypotheses. We describe strategies for setting up protocols and determining what type of experimental conditions are necessary for a valid experiment. We

Figure 15.10 A book prospectus. (*Continued*)

define qualitative and quantitative research and describe some typical methodologies associated with each. We also outline the research process, from gathering ideas for a new experiment through recording and analyzing results.

8. Informational Interviews on the Job

A Four-Step Approach to Interviews
Multiple Interviews with Different SMEs
Conference Calls
E-mail Interviews
Teleconferencing
Computer Conferencing
Summary
Research Activities
For Further Reading

9. Interviews for Empirical Research

Interviewing Skills
Rapport with Subjects
Confidentiality
Locations for Interviews
Conditions under which Interviews Take Place
Time Restrictions
Follow-up Interviews
Summary
Research Activities
For Further Reading

Researchers who record responses to in-person questionnaires, interview respondents, and work with a limited number of people to study them in some way in depth need to develop special skills to gain their interviewees' confidence and record accurately the information they gather. In this chapter we emphasize ethical behavior and good research practices in gathering information from subjects, gathering information for formal research through interviews, and ensuring that the information is usable.

10. Questionnaires and Surveys

Advantages and Disadvantages of Questionnaires
Types of Questionnaires
Variables to Consider in Designing a Questionnaire
Types of Questions
Question Sequences
Wording of Questions

(continues)

Figure 15.10 (*Continued*)

A Sample Questionnaire
Methods of Distributing Questionnaires
Methods of Collecting Questionnaires
Following Up on Questionnaires
Response Rates
Resulting Data
Summary
Research Activities
For Further Reading

The practical aspects of creating, disseminating, collecting, and following up on questionnaires are discussed. As well, we provide sample questionnaires and analyses of the language and format used in each.

11. Letters of Inquiry

Purpose of the Letter
Benefits of a Letter
A Request for Advertised Materials
An Unsolicited Request for Information
Summary
Research Activities
For Further Reading

We describe situations in which letters of inquiry are a valuable source of information and provide examples of letters that can elicit the desired response.

Part III. Ways to Evaluate Results

12. Evaluation of Secondary Sources

Authority of the Source
Credentials
Dated Information
Quality of the Information
Style of the Information
Conflicting Sources
Summary
Research Activities
For Further Reading

When researchers review secondary sources, especially for background information or for literature reviews, they need to evaluate the quality of the sources and have a method of dealing with discrepancies in information. In this chapter we discuss ways

Figure 15.10 A book prospectus. (*Continued*)

that researchers can evaluate and rank the usefulness of secondary sources for their type of research project or interest.

13. Statistics and Measurements

Statistical Terms
Developing the Right Test for Your Results
The *t* Statistic
Analysis of Variance (ANOVA)
Measures of Analysis of Variance (MANOVA)
Regression Analysis
Log-linear Analysis
The Chi-square Statistic
Other Statistical Measures
Summary
Research Activities
For Further Reading

In this chapter we provide an overview of the vocabulary used in statistical analyses. We describe various statistical tests that can be used to interpret data and help technical communicators and other researchers understand how to interpret results. This chapter is not designed to be an in-depth statistics textbook or manual, but to provide researchers with a basic understanding of the types of analyses they might apply to their data and to work with experts who are well-versed in statistics.

14. Usability Tests and Validation

Usability Tests of Technical Documentation
Validation of Technical Information
Development of a Usability Test
A Sample Usability Test
Summary
Research Activities
For Further Reading

We define validation and usability testing and describe how technical communicators can develop ways to test a single document or to help establish ongoing usability testing of multiple documents produced by a company.

Part IV. Possible Publications of Results

15. Prospectuses

Query Letters
Creation of a Prospectus

(continues)

Figure 15.10 **(*Continued*)**

A Sample Prospectus
Submission of a Prospectus
Follow-up Activities
Summary
Research Activities
For Further Reading

Because the results of research should be shared, we offer this chapter of sample query letters and prospectuses to help researchers locate appropriate publishers for their research. This chapter should also provide a practical guide to describing research and possible publications.

16. Abstracts and Articles

Abstracts for Journals and Conference Proceedings
Conference Papers
Journal Articles
Chapters in Reference Works
Summary
Research Activities
For Further Reading

We emphasize ways to write an effective abstract according to a publisher's specifications. In addition, we describe different types of professional publications found in technical communication, technology, and the sciences and the kinds of information that are included in each. We also briefly describe the process of submission, review, and revision before publication.

17. Bibliographies

Publishable Bibliographies
Database Bibliographic Entries
Journal Bibliographies
Citation Styles for Hardcopy and Softcopy Sources
Summary
Research Activities
For Further Reading

We discuss different formats for bibliographies and types of publications that print bibliographies. We explain ways to annotate entries and submit bibliographies for possible publication.

18. Multiple Publications from Research

Ethical Use of Research
Multiple Submissions

Figure 15.10 A book prospectus. (*Continued*)

Possibilities for Multiple Publications
Ongoing Research Interests
Summary
Research Activities
For Further Reading

The ethical way to submit information for one or more publications is emphasized in this chapter. We describe accepted practices for submitting drafts for possible publication and ways to write more than one publication based on one's recent research.

19. Funding for Further Research

Research Grants
Business and Technological Research
Research Institutes
Academic Research
Summary
Research Activities
For Further Reading

We define research grants and provide an overview of sources where researchers may find grantors for additional or new research. We briefly illustrate how a grant proposal is written.

Appendix A. Selected Secondary Sources for Technical Communicators

Appendix B. Checklist for Researchers for In-house Publications

Appendix C. Checklist for Researchers Seeking Funding

Glossary

Index

Figure 15.10 ***(Continued)***

Summary

When you have a short publication about which you need to query a publisher, you send only a query letter, unless an outline or a sample is requested. You should know the publication and the editors' preferences regarding query letters, including what they must contain, before you write. Sometimes the query can be rather informal, such as a phone call or an e-mail message. But most often, an official query letter needs to be in print and in letter form.

When you want to submit a longer work for publication, you may need to develop a proposal letter that can be sent by itself or possibly serve as a cover letter for accompanying materials, like a sample chapter or section, an outline, or a description of each chapter or section. If you're proposing a book-length document, you need to submit a prospectus. You might need a proposal letter to preface the packet of materials, but you frequently need only a letter of transmittal to introduce the prospectus. Then the outlines, samples, and supporting documentation can flesh out the prospectus.

Remember: If you think an idea is worthy of publication, take the time to write an accurate, interesting query, proposal, or prospectus. Part of your job as a researcher is to find ways to tell others about your research, and to do that, you often have to sell your idea to a publisher.

Research Activities

1. Complete the background research about at least one publisher who would be interested in your current project.
   - Does the publisher require a query letter?
   - Does the publisher require a proposal or prospectus?
   - Who is the contact person who should receive your materials?
   - How often is the publication published?
   - What are the specifications for publication?
   - Why does this publisher seem like the best one for your information?
2. Send a query letter to a journal and suggest an article based on your research project.
3. Send a proposal to a professional conference specializing in your area of research expertise.
4. Submit a prospectus for a book based on the findings of your research.
5. Begin a file with information about and received from publishers. You may want to develop online or disk-based files to store your

notes, and you may be able to download information from networks about certain publishers. Keep hardcopy and softcopy records of the information you've gathered about publishers, as well as electronic and hard copies of the materials you've submitted for publication or presentation.

For Further Reading

Lovell, Ron. 1988. "How to Write Textbooks for Fun, If Not for Profit." *Journalism Educator* (Fall): 34-5.

Moxley, Joseph M. (Ed.). 1992. *Writing and Publishing for Academic Authors.* Lanham, MD: University Press of America.

CHAPTER SIXTEEN

Abstracts

For most of us, it's probably a real source of pleasure to browse through the various titles at a bookstore, reading the excerpts on book covers trying to decide what we'd like to buy to entertain ourselves. However, when we're on the job with very specific deadlines for producing thoroughly researched manuscripts, we don't have that kind of time.

For on-the-job research purposes, we want navigational and informational tools that help us quickly locate all the information we need about a particular subject matter. Once we've located all that material, we want the quickest and most effective way of focusing our attention on those materials that provide us the best information. Among other purposes, abstracts serve as those navigational and informational tools.

Two kinds of abstracts with which you need to be most familiar are *topical* (also called descriptive or indicative) *abstracts* and *informational abstracts*. In this chapter, you'll read about both types of abstracts.

Topical Abstracts

Topical abstracts are brief descriptions of a document. They are designed to tell potential readers about the document, for example, what kind of manuscript it is, what kind of information it's designed to provide, and

In this paper we describe the special planning needs of nontraditional students who are changing careers, entering the job market for the first time, or reentering the job market. These nontraditional job seekers' psychological, emotional, and logistic considerations are discussed. We provide some advice for those nontraditional students seeking changes in their careers.

Figure 16.1 A topical abstract.

how the information is organized. Topical abstracts do not summarize the document's content. Good topical abstracts help you decide if you need or want to read the document, and if the type of information will be useful. Figure 16.1 is a topical abstract.

Note what you learn from the topical abstract in this figure. You learn that:

1. The document is a paper.
2. In the paper, the authors "describe." This verb is very different from *report*, *state*, or *include*, for example.
3. The authors describe "special planning needs of nontraditional students who are changing careers, entering the job market for the first time, or reentering the job market." Now you have an idea of the subject and the number of topics that are covered in the paper.
4. The authors also "discuss," which indicates a more thorough treatment of psychological, emotional, and logistic considerations.
5. The authors "provide some advice."
6. The authors believe that their audience is primarily made up of nontraditional students, as indicated by the last line in the abstract. For whom did they provide the advice? Nontraditional students. If you're part of that group, you might find the information especially useful.

Note, too, what you *don't* learn from this topical abstract. You don't learn

1. What those "special planning needs of nontraditional students" are.
2. What the "psychological, emotional, and logistic considerations" are.
3. What the "advice" is.
4. By implication, what the authors' attitudes, ideas, or perspectives are about any of these topics.

If you're looking for information designed to note some special planning considerations for nontraditional students, or information that provides them advice, you add this paper to your list of materials to gather for your research. If you're looking for anything else, then you eliminate this paper from your reading list.

The topical abstract, then, is designed to help you complete both initial gathering and elimination of information you may need to complete research for a particular topic. They can aid you in these ways:

- When you're initially gathering materials, they help you select materials to add to your reading list.
- If you have materials in hand and need to decide which ones you want to read, topical abstracts can help you decide which to read and which to eliminate; or they can help you prioritize your materials for reading.
- Topical abstracts can also be useful if you have a paper filing system with limited space, because you can save space by filing only one page of a document—the topical abstract. The rest of the document can be stored elsewhere. If you're using an online system, you can easily store topical abstracts in a database, so that you can retrieve them by searching for keywords in the title or the abstract itself.

Informative Abstracts

Informative abstracts, on the other hand, are designed to provide specific information. These abstracts contain the primary information discussed within the text they abstract. They summarize the content, instead of simply describing what kind of information can be found in the document. For this reason, you find within an informative abstract a specific discussion of the research problem handled, the procedure used to conduct the research, any significant findings or conclusions, and any recommendations or suggestions for further research. This type of abstract provides a report in miniature that includes information from every major section in the document.

Informative abstracts contain the essential information, without elaboration or description, that you find in the entire document. If you are in a situation in which you need to know quickly the most important information, without a detailed discussion, you rely on the abstract to provide that information.

Nontraditional students may need to take a technical writing course for several reasons: they want a promotion; their supervisor assigns them new job responsibilities that include technical writing; they have poor communication skills; they are changing careers; they want to move into a marketable field like technical communication. However, nontraditional students have special educational needs and expectations, and educators must consider the technical writing course's structure, the method of presenting information, the range of students' learning styles, and assignments to meet nontraditional students' career needs and academic expectations. As the number of adult learners increases in the next decade, and as technical communicators need to receive additional education, current educational practices must be reevaluated. Nontraditional students need to learn how to approach educators when their needs are not being met and how to tailor resumes and portfolios to reflect their life experiences, previous work experiences, and current education.

Figure 16.2 An informative abstract.

Sometimes the informative abstract has another name, depending upon the type of document being summarized. An *executive summary*, for example, is used in business reports to give busy executives the information they need to make decisions when they don't have time to read all the paperwork that crosses their desks. The executive summary is traditionally used in feasibility reports or other business reports, for example. Figure 16.2 is an informative abstract; the topical abstract shown in Figure 16.1 was written for the same paper.

This abstract summarizes

1. Specific reasons why nontraditional students may take technical writing courses.
2. Specific indications that nontraditional students have special needs and expectations.
3. Ways that educators have to adapt their materials and methods for nontraditional students.
4. What nontraditional students need to do as they take classes and change careers.

This abstract doesn't provide any kind of description of the document. It emphasizes the subject of the document.

Figure 16.3 is another informative abstract from a technical article published in a psychology journal. The abstracts you've read earlier in this chapter were written about a paper, like those given at a conference presentation, for a more general readership. This abstract was written for

Guided by descriptions of the traditional sexual script, researchers have examined men's, but not women's, attempts to influence a reluctant dating partner to engage in sex. In the current study, we examined both men's and women's experiences of women's attempts to influence a reluctant male dating partner to engage in sexual activity. Ninety male and 111 female participants completed a series of questionnaires assessing sexual/dating histories and attitudes toward sexuality. Participants also completed the Sexual Situation Questionnaire designed to assess the situational and behavioral characteristics of disagreements in which a woman attempted to influence a reluctant man to engage in sex. Fifty-six percent of the participants reported having experienced such a disagreement interaction in the year prior to the study. In accordance with attribution theory, men were expected to be more likely than were women to make personal attributions for their reluctance, whereas women were expected to be more likely than were men to make situational attributions for men's refusals. Our findings supported these predictions. Most participants indicated that the woman complied with the man's indication of reluctance and used no influence behaviors. Somewhat fewer indicated that the woman complied but also used some form of influence, whereas a distinct minority indicated that the woman did not comply with the man's refusal. No influence strategy was rated consistently as having had a positive, negative, or neutral impact on the man, which indicates that influence behaviors can have varying impacts depending on the context in which the influence occurs. In general, few negative emotional or relational consequences were found to be associated with these disagreements. As most disagreements did not result in engaging in the disputed level of sexual activity, women appear to have had little success in their attempts to orchestrate sexual encounters (traditional men's role in sexual dating interactions).

Reprinted from "Eroding Stereotypes: College Women's Attempts to Influence Reluctant Male Sexual Partners," Lucia F. O'Sullivan and E. Sandra Byers, *The Journal of Sex Research*, Vol. 30, No. 3, p. 270, August 1993, by permission of the Society for the Scientific Study of Sex.

Figure 16.3 Informative abstract from a scientific article.

readers with specific interests in specialized areas of psychology, people who are very familiar with qualitative and quantitative research interests.

This abstract is based on a scientific method of information gathering and is intended to provide much more detailed information about the data collected. Thus, this abstract provides

1. the subject of the research
2. the reason for or background of the research
3. the research methods
4. the results
5. some conclusions from the research

And depending upon the type of research and the document being abstracted, an informative abstract might also include some recommendations for further activity.

As a reader of this abstract, you know the significance of the data. Specific numbers (90 male and 111 female participants, and 56 percent of participants) are listed, so you have a good idea of the scope or possible limitations of the study. If you are specifically interested in what questions the respondents were asked, how the questions were asked, who asked them, and during what time period, for example, you can read the article. But the abstract provides the essence of the information.

This type of abstract is helpful when you first determine if the content is useful to you, but it is also important after you've read the article. As you conduct research, you'll find that informative abstracts will help you refresh your memory about the importance of a study. After you read several sources, you may not recall all the details of individual articles or books, for example, even when you take extensive notes. An informative abstract succinctly summarizes the important details you'll need to know.

Locations of Abstracts

Abstracts generally appear in four places:

- in abstract journals or collections
- with library citations (for example, databases, indexes, or card catalogs)
- preceding many scholarly articles (for example, journals and conference proceedings)
- with prefatory matter in longer, more formal documents (for example, formal reports)

Abstract journals like *Dissertation Abstracts* provide a service for researchers who need to keep up to date on research topics, researchers conducting particular types of studies, and the numbers and types of current projects. Well-known abstract journals within a discipline such as chemistry, for example, can save you hours of work skimming sources. Some abstract journals indicate works that have not been published, or works that will soon be published, such as dissertations and theses. Without using abstract journals, you would have little idea of the kind of research going on at any one time.

It's only logical, too, that abstracts, or at least brief annotations, appear in library citations. Many computerized databases provide abstracts with the bibliographic information to save researchers time in finding sources. Also, sources that are hard to locate may be considered when researchers read the abstracts; based on the abstracts, they can decide whether to order the source.

Finally, scholarly articles and papers are usually prefaced with an abstract. Writers submitting articles to be published in journals and conference proceedings are usually required to include at least one abstract, and sometimes both types of abstracts are required.

Informative abstracts are necessarily longer than topical abstracts, and, thus, usually appear on a separate page in a long document, such as a formal report, or at the top of the first page of an article or a conference paper published in conference proceedings. But informative abstracts can also be found in collections of abstracts that document the amount and types of research conducted within a profession or a discipline during one year or over several years.

Topical abstracts are usually very short—just a few sentences at most. They often fit along the bottom of the title page in reports, in program descriptions of conference presentations, or in other introductory materials that highlight the topics to be covered in a presentation, an article, a report, or another document. Some may appear on a copyright page or even on a separate page at the beginning of a manuscript. Wherever they occur, they should be clearly identified as topical abstracts and be easy to locate, so that readers can quickly determine if the type of information in the document will help them.

Abstracts That You Write

You usually write abstracts on two different occasions: First, you might abstract materials you read in your research. In this case, the abstract is for your use only and is focused toward your needs. An abstract for your use might contain the following:

- The title and possibly bibliographical information about the document you're abstracting.
- Comments regarding general areas of support for the thesis.
- The thesis of material *directly* related to your interest (remember that frequently only parts, or even a single part of a text you read, will relate directly to the research you're doing).

- Major supporting points of the thesis related to your subject.
- Personal comments about the reliability of the information.
- The source's importance to your particular research.

The result of providing these kinds of information resembles in many ways an annotation. However, it differs in that it is *informative* rather than *descriptive.* Your abstract will be your source of information for making arguments, citing references, using the material to replicate an experiment you wish to do, and ensuring accuracy in your reporting.

The second occasion when you may need to write an abstract is if you are submitting something you have written for publication or presentation, perhaps at a conference. Whether you write a topical or an informative abstract, and what you put into each, depends on the publisher's requirements or specifications. For example, if you were giving a presentation at a conference that published proceedings of papers submitted, you might be required to submit an abstract first to be reviewed by a committee who selects speakers for the conference. In such a case, you might find yourself writing a multipage informative abstract.

Once your article or presentation is accepted, you might be asked to write a very brief (perhaps 50- or 100-word) abstract that clearly tells potential readers or conference attendees about your subject and does so in language that attracts their attention.

Journals that publish your article, or abstracting services that publish the abstract of your article or a special document, such as a dissertation or a thesis, may require you to submit an informative abstract. You're often given a word or character limit, and within that limit, you have to summarize the content in clear, precise, and interesting language. Many times you're also asked to provide a list of keywords found in your abstract, so that online databases containing your abstract can be searched by keywords.

Although these publication and presentation scenarios are common to most technical communicators' work, you might find yourself in special situations that require you to write an abstract. For example, if you're writing a project proposal in an effort to obtain a grant, frequently you have to submit an abstract of the proposed project, along with a full description and other supporting evidence of your qualifications to complete the project. The abstract may have to conform to space limitations provided on a form or to specifications on the number of characters or words. Nevertheless, your informative or topical abstract is an important part of the whole package, and often this brief abstract is the first item readers want to see.

General Suggestions for Writing an Abstract

Summarizing a whole document into a few words may seem like an impossible task. But it's easier if you write the abstract after you've written everything else in the document and when you've reviewed the document to make sure you're familiar with all of it. The following suggestions, however, should help you write an abstract for any document:

- Read the document's title and then the summary.
- Read the statement of purpose and the results or conclusions.
- Read the document for keywords.
- Read the document selectively, looking for the main ideas described in the title, summary, statement of purpose, conclusions, and keywords. Take notes as you read.
- Write the abstract with an intentional sequence, which may not be the organization of the document. For example, you might want to write the abstract following a why, what, how, what results pattern.

This process can help you write abstracts not only of your own documents but also for other documents.

Both descriptive and informative abstracts have length and content requirements, but they also have style requirements that are just as important as the content requirements. Abstracts should always be written in clear, direct, and precise prose. You should meet space limitations through careful consideration of content, not through the elimination of words like the articles (a, an, the), or through qualifying words that alter the accuracy of statements.

As a writer of abstracts that others will read, you are professionally and ethically obligated to ensure that the abstracts are accurate representations of the documents being abstracted. As a writer of abstracts you alone will use, you ensure that you can refer to them for accurate information.

Suggestions for Writing Topical Abstracts

When you write a topical abstract, do the following:

- Identify the type of work you're abstracting (for example, paper, manual, book, presentation).
- Indicate how the authors handled the material (for example, do they describe, discuss, analyze, define?).
- Note the general subject area.
- Note in ending whether the material provides suggestions, conclusions, recommendations, and so on.

Also, you should keep the abstract short and precise. Topical abstracts should never be more than 100 words long and usually are only about 50 words.

Suggestions for Writing Informative Abstracts

Because informative abstracts contain more specific information, it is generally better to ask specific questions about the content and use the answers to these questions as the basis of the content of the informative abstract. Here are some suggested questions:

1. What is the specific research problem?
2. What is the history (background and significance) of the subject?
3. What are the benefits of solving the problem or studying this subject?
4. What procedures were used to solve the problem or conduct the research?
5. What were the results? Were they statistical? numerical? descriptive?
6. What conclusions were drawn?
7. What recommendations for action, including further research, were provided?

When you create the abstract, write one specific sentence to answer each question. Choose modifiers carefully so that you precisely and accurately describe the research or project. The informative abstract must reflect accurately, including the proper emphasis, the content of the text it abstracts, because some audiences will depend upon it alone to provide them information on which to make decisions.

Standards and Guidelines for Abstracting

Many national and international institutions, organizations, and committees are concerned with standardizing abstracts. You should know the standards for writing in your field. Style manuals usually include a section illustrating the type of abstract you should use for publications in a specific discipline. For example, most journals list the type of, and any specifications for, the abstract you're required to submit for publication. Committees reviewing grant proposals or conference-presentation proposals also provide guidelines in their solicitation notices. For specific guidelines, you should refer to the publication, solicitation notice, or publisher.

If you're unsure about the way to write abstracts or need more information about standards in abstracting, you can check catalogs provided by the National Standards Institution, which frequently contain recommendations from the International Organization for Standardization. International standards are recognized by the letters ISO, followed by a serial number. Draft International Standards have the prefix DIS, followed by a number; British Standards use the code BS; Indian Standards use IS; and American Standards use ANSI. You'll find one reference for an ANSI standard listed in the additional readings section in this chapter.

Summary

Topical and informative abstracts often accompany scholarly documents in publication and may be found in descriptive works like library card catalogs, indexes, and abstract journals. Being able to write a good abstract requires you to summarize major points and organize your summary into a logical description that accurately reflects the whole document. Whether you write abstracts of your own documents for publication, read abstracts to help you keep up with new information, use abstracts to help you choose sources for your research, or write abstracts to help you keep track of the sources you read, your professional development, to a great extent, will depend on your ability to work with abstracts.

Research Activities

1. Write an informative abstract of a research document you're currently creating or have recently written, following the process outlined in this chapter.
2. Write abstracts of the secondary sources you used in your initial research for your current project. Keep these abstracts on file, either online or on paper, for future research needs.
3. Study the abstracts to articles in several different journals in your field. Then read the publication specifications for abstracting articles.
4. Review a recent issue of an abstract journal to learn trends in research in your field.

For Further Reading

Achterberg, Cheryl. 1989. "Tips for Writing Abstracts." *Journal of Nutrition Education* 21(5) (September): 197.

American National Standards Institute. 1979. *American National Standard for Writing Abstracts. Secretariat: Council of National Library and Information Associations.* (Standard Z39.1401979, a revision of Z39.14-1971.) New York: The Institute.

Cremmins, Edward T. 1982. *The Art of Abstracting.* Philadelphia: ISI Press.

Moxley, Joseph Michael. 1992. *Publish, Don't Perish: The Scholar's Guide to Academic Writing and Publishing.* Westport, CT: Greenwood Press.

CHAPTER SEVENTEEN

Bibliographies and the Documentation of Sources

Bibliographies are lists of sources that provide complete citations of references authors use when they conduct research. Although brief documentation of sources being quoted, summarized, or paraphrased may occur in the text, the bibliography provides all the details readers need to locate the sources. Bibliographies can indicate which sources were directly used in the research, which were consulted as background material, and which were not used in the research but might be useful additional sources. Depending upon the style manual the authors used, the bibliography may be called a List of Works Consulted, Works Cited, or References.

Most of us are probably familiar with bibliographies at the end of research documents, such as articles, books, or reports. But other bibliographies are also important at different parts of the research process. For example, a published bibliography has no accompanying article or book. The bibliography itself is the publication. Usually the citations listed in the bibliography are annotated, so that readers have a synopsis of the source and its importance, as well as enough information to locate the source. However, a bibliography may just list hundreds of sources, perhaps categorized under subheadings.

Although bibliographies are an important part of the way sources are documented, other textual elements can also be used to comment on, keep track of, and help you locate sources. In-text citations, notes, footnotes, and endnotes are also useful to help you (and your readers) learn more about a source and its importance in the research process or product.

You may find sources cited within information you find online, but the bibliographic citations and supporting documentation are usually written as text. For example, an article from conference proceedings may be presented online, so that text and graphics may be used by anyone who has access to this electronic source. Within the article, just as within a paper document, you may see in-text references, bibliographic citations, and notes. You may locate additional information about the source or other secondary sources the author feels are important by using hypertext links or scrolling through the text until you come to a bibliography. But the bibliographic documentation is most likely to appear in prose. That's why the emphasis in this chapter is on textual documentation. The same type of information is likely to be found in prose, whether in a hardcopy document or softcopy information.

Preliminary Bibliography

The term *preliminary bibliography* is actually used to describe two different kinds of bibliographies. One is a list of documents you're consulting or plan to consult during the early stages of a long-term project. This type of bibliography is useful just to you, because it presents a sampling of the available sources and generally lists only a few works. The bibliography may be very informal; usually it conforms to whatever shorthand style is easiest for you to use as you locate sources and make comments about them. This bibliography begins when you write your first potential source of information and continues to grow as you find various sources that you might use as your research continues.

A second type of preliminary bibliography is one written for others to use. Although the bibliography is still preliminary, because more sources will be added, it looks more formal than a bibliography you've created just for your use. You may put this type of bibliography in a formal citation style, such as that recommended by the Modern Language Association (MLA) or the American Psychological Association (APA) in their style manuals.

A Preliminary Bibliography for Your Use Only

A preliminary bibliography can be started as soon as you begin your first general search for sources. For example, if your library has electronic access to databases or an online catalog, you might begin by typing some keywords about your subject matter into the computer system. Those keywords lead you to a list of sources. From that list, you collect some names of authors or titles that look promising. You then type information about each of these sources into the library's computer system to get information about that source. For each book-type source, you receive standard library information: author, title, publisher, place of publication, year of publication, ISBN or ISSN reference, and the number/letter code that locates the work within your library or within another library (in the case of interlibrary loan or interlibrary databases). Usually you also get a brief description of what the work contains. In some online databases, you might find more than a brief description; you may also be able to read the table of contents or an index of key topics. Whether you transfer the citations to note cards, print them directly from the computer, or transfer them to a personal computer file, you've begun to create a preliminary bibliography.

As you look through these sources individually, you make decisions about whether you should look at the particular sources further or disregard them. When you decide to look at a source, you probably should make some notes about it. Because this bibliography is for your own search purposes, you may write down no more information than the source's author, title, and location, but you've begun a bibliography. For example, if you were looking for books about research in technical communication, you would probably find this book listed. On a reference card or on a computer file you might write only the following information:

> Porter, Lynnette R., and Coggin, William O. *Research Strategies in Technical Communication,* and the Library of Congress number that gives the book's shelf location

You might list several sources that seem appropriate and interesting for your research long before you ever track down the sources and decide if your initial perception about the source's usefulness is correct.

After you've found the sources that you think will be useful, you decide which ones you want to read. Then your bibliography begins to take on characteristics to help you keep track of your research. For example, to your original listing in the preliminary bibliography, you might add details like the publisher's name and the publication date. Or you might decide to provide some short, descriptive notes about the book. That is, you can annotate the reference; you add information about the source that will remind you of the source's contents or their potential value to you. (See the upcoming section about annotated bibliographies for examples.)

A Preliminary Bibliography for Other's Use

A preliminary bibliography also may appear in a proposal or progress report. You may need to submit this kind of bibliography if you're proposing a conference paper, a research plan, or a project to be funded by your group or department. In these instances, the preliminary bibliography tells your reader that you've consulted certain works to this point in your research, and you plan to consult others as you complete the research.

A preliminary bibliography in a manuscript that others will read contains complete citations so that your audience has all the information they need to verify the sources. The bibliography also provides readers with an overview of the number, type, and recency of available sources or consulted sources. Finally, this type of bibliography can indicate your diligence as a researcher. If you've missed important sources that every knowledgeable researcher is expected to know, or if your sources are clearly out of date, then your credibility or status as a researcher in this field may be questioned. If you've covered a broad range of important topics within an appropriate range of publication dates, you are given more weight as a researcher, and your project is more likely to be approved. Although your readers understand that the bibliography will grow or change as you conduct more research, the preliminary bibliography is a good measure of the quality of the work to come.

Evaluating the Sources Listed in a Preliminary Bibliography

Whether you've listed sources to look up for your own use or have indicated to readers that you plan to use those or similar sources as you conduct further research, you and your readers can learn a lot about the sources just by looking at the preliminary bibliography.

The items in Figure 17.1, for example, might appear in a preliminary bibliography for an article about Standard General Markup Language (SGML). (The sources have been put in a consistent format, but they may

Ahearn, Hally. (May 1993). SGML and the *New Yorker* magazine. *Technical Communication*, 40(2), 226-229.

Alschuler, Liora. (May 1993). Special section: Standard general markup language. *Technical Communication*, 40(2), 208-209.

Andrews, Dave. (March 1993). Portable documentation accelerating SGML. *Byte*, 18(3), 32.

Goldfarb, Charles F. (1990). *The SGML handbook.* Ed. Yuri Rubinsky. New York: Oxford. Library of Congress number (Ref) Z286.E43G64 1990.

Raggett, David. (November 1994). A review of the HTML + document format. *Computer Networks & ISDN Systems*, 27(2), 135-145.

Figure 17.1 A short bibliography in APA style.

need to be edited so that the format conforms to whichever standard style manual is eventually required for publication or submission.)

From the titles of the sources in the sample bibliography, you can see that they all describe markup languages, including SGML and Hypertext Markup Language (HTML), but you don't know the focus of each source. Words like *review* indicate a summary article, whereas *handbook* suggests a practical, comprehensive guide. The name of the publication or publisher can also provide clues about the content of the source; *Computer Networks & ISDN Systems* would seem to have a more technical readership than *Byte,* but if you're not an expert in systems, you might find the language and number of definitions initially more useful in *Byte.* If you know that *Technical Communication* is the journal of the Society for Technical Communication, and you see a special section or several articles within the same journal, you may also get a sense of the importance of this topic to a particular group, in this case, technical communicators. By looking at the titles of the documents and the publications, you can better determine if the information is specific to your project, or if the source should be eliminated as a possible source of information for the project.

Familiarity with other works by the authors may help you decide if the information will be useful to you. If you know that Charles Goldfarb regularly publishes works about SGML, and his work has been well received, you can determine if his current article is one you must read. When you look at the author's name, you should note whether the author wrote the book, article, paper, or whatever; only presented it; or edited or compiled works by other authors. Yuri Rubinsky is listed as an editor; the source therefore may contain articles or chapters from other authors selected by Rubinsky for inclusion in one volume, or Rubinsky may have significantly edited the materials provided by Goldfarb. You might want to check this work to see who wrote the chapters, how recent the original information is, and how much influence the editor had in the published work.

In your evaluation of the preliminary bibliography, you need to know how much information has been recently published, and if the information indicates trends in recent research and publication about your topic. If this sample list represented all the sources you could find about SGML, you could see some trends in the scholarship about that subject. For example, you might find that not much was written in the late 1980s, but a significant number of articles and a lesser number of books were published beginning in the early 1990s. If you're interested in other markup languages in addition to SGML, you might be interested to see more recent publications about HTML. By looking at the publication dates, amount of information, and the number and type of publications, you can get a better sense of trends and a good sense of the scope of information about a topic.

Annotated Bibliography

In some instances you need more information about a particular source than just reference notes. For example, as you select material to read, you may add a brief description of the work. Sometimes you add these notes when you first write down a source's title; you might add descriptive notes from information listed in the card catalog or an annotation provided with a database listing of the source. Other times you will add descriptions after you've read a work and can make some value judgments about the work's usefulness to a particular research project. In either case, when you add descriptive information about the source, you've begun to annotate the bibliography.

An *annotated bibliography* is a list of sources that contains short descriptions of each work cited. An *annotation* consists of critical and evaluative statements about the source. In the annotation, you assess the source's reliability and discuss its significance to and importance for the potential user. You might write an annotation to discuss the author's credentials, the work's subject and purpose, its format, and any errors, weaknesses, or strengths. Some annotations are written in full sentences, whereas others are written only in concise phrases. Whichever format you choose, however, all your annotations should be consistent.

Because of the amount of material published, it is very difficult for researchers to read everything available on a given subject. To help determine both what kinds of material are available and the possible value of those materials for any given research project, some annotated bibliographies are published as separate manuscripts. Some appear as articles in professional journals, whereas others appear as separate books. These annotated bibliographies can be particularly helpful to you as a researcher when you're first gathering information. The annotation might tell you that a source is outdated or does not contain the focus you need to research; therefore, by reading the annotation, you decide to skip one source in favor of another that can provide you with information directly related to your project. Figure 17.2 is a sample page from such an annotated bibliography.

Literature Review

Technical communicators must sometimes show first in their manuscripts that they have a grasp of the existing literature on a particular subject before they can begin to discuss new ideas or to describe the focus for their original research. In books, articles, theses, dissertations, or other manuscripts in which you claim to be expressing original ideas, you have to show what others have said, written, or researched about the subject so that when you make your points, you can point out how they differ from or build on others' work. In other situations, you may need to describe what you've read so that you can refute some ideas and recommend others for your readers. In either case, when you include discussions about the works you've read, you're creating a literature review.

A literature review generally is among the first parts of a manuscript you write. In a book, thesis, dissertation, or scholarly article, for example, it may be presented as one of the first chapters or sections of the published work. A literature review is a kind of focused annotation about a work. Within the review, you indicate the work and the author's/researcher's thesis. You might discuss the author's/researcher's methodology, materials, participants, results, analyses of data, and/or recommendations. You can then add evaluative evidence either to show your support for the previous work or to indicate how your work improves upon, builds on,

Reprinted bibliography from the *Bulletin of the ABC*.

The Bulletin, September 1990, Page 36

ANNOTATED BIBLIOGRAPHY FOR TEACHING ETHICS IN PROFESSIONAL WRITING

Bruce W. Speck
Memphis State University

Lynnette R. Porter
The University of Findlay

How and whether to teach ethics is hotly debated in education today. Some professionals contend that teachers should not impose ethical values on students; others say that the teacher simply cannot teach value clarification, which, they claim, is a moral wasteland. A third group says teachers shouldn't teach ethics at all. This bibliography aims to help teachers solve the problem for themselves by providing them with a wide variety of sources about teaching ethics in the professional writing class. However, it wholeheartedly recommends, by its very existence, that teachers of professional writing incorporate instruction in ethics into their classes.

To present an overview of available works that discuss ethics in professional writing, we have annotated books, articles, and conference papers to provide

(1) a perspective on the study of ethics in professional writing

(2) a reading list for educators, students, and businesspeople

(3) materials (case studies, for example) educators might integrate into their courses.

The bibliography is divided into the following areas:

Theory Focus

Teaching Focus (including subsections on General Composition, Journalism, Speech, Business Writing, and Technical Writing)

Teacher Focus

Related Bibliographies

These four sections comprise three important foci for educators who teach students how to develop an understanding of ethical issues. The first focus, theory, emphasizes works dealing with studies, hypotheses, and methods of measuring ethics. The second focus, teaching, emphasizes practical applications of research and theory in classroom activities. In that section we describe articles which provide suggestions for lessons, courses, and curriculum design. The third focus, teacher, emphasizes the educator as role model, communications professional, and student of ethics. We believe that all three foci are necessary for teachers to understand ethical issues and develop materials to improve the quality of ethics education in professional writing. The concluding Related Bibliographies section cites other sources educators can consult to integrate ethics education in their professional writing courses.

In annotating the bibliography, we have refrained from evaluating sources. Rather, we have attempted to describe the substance of each source so that other researchers can make an informed decision about whether to read a source in its entirety. Certainly, our efforts are flawed by the tedious problem of imperfect human perception, but we trust that any author who believes that we have misrepresented his or her work—however slightly—will accept our expressed intention to be fair-minded. Indeed, we trust the intention to be fair-minded will be apparent throughout the bibliography.

ANNOTATED BIBLIOGRAPHY FOR TEACHING ETHICS IN PROFESSIONAL WRITING

THEORY FOCUS

1. Bahm, A. J. (1982). Teaching ethics without ethics to teach. *Journal of Business Ethics, 1*(1), 43-47.

A rebirth in the teaching of ethics focuses on applied ethics, not questions about "some ultimate bases accepted as compelling consent for convincing reasons" (43). Such a view of teaching ethics begs the question of what ethics one should teach. The author questions the legitimacy of American culture's attempt to find answers to urgent and complex ethical problems without consulting ethical theory.

2. Beauchamp, T. L., & Bowie, N.E., Eds. (1979). *Ethical theory and business.* Englewood Cliffs, NJ: Prentice-Hall.

This anthology of articles was created as a textbook to promote discussion of controversial issues. The articles were selected to reflect a combination of moral philosophy and business, and to indicate many different sides of controversial issues in a debate format. The book covers business applications and theories of ethics; it includes information about new theories of corporate responsibility, codes of ethics, and conflict of interest. As a broad overview of ethics and business, it discusses ethics in advertising, hiring practices, and reverse discrimination.

Figure 17.2 A published bibliography.

or otherwise differs from the original work. Works discussed in a literature review are almost always given complete reference citations in either a note, such as an endnote or a footnote, or a bibliographic citation.

The literature review establishes your credibility as a researcher in three ways:

1. It illustrates your understanding of relevant research that preceded yours and familiarity with important researchers and works in your field.
2. It provides a solid foundation on which your original research rests and provides evidence of the soundness of your ideas, methods, and interpretations.
3. It provides a practical review of important works and ideas underpinning your research and also offers (through complete citations in notes or a bibliography) a list of sources to which your readers can refer.

The excerpt shown in Figure 17.3 from a literature review was written for a research article in a psychology journal. As is typical with many literature reviews, this section is long in proportion to the rest of the manuscript. (For example, the literature review chapter in a dissertation or a thesis may be around one-fourth or one-fifth of the entire document. In a 10-page article, the literature review may be two or three pages.) Although in some journals this section may be titled "Literature Review," in this journal the literature review is incorporated into the introduction and first section of the article.

The complete reference citations for the documents cited in this literature review appeared in the References section at the end of the article. This article was published on 13 full (8-1/2 x 11-inch) pages, which included the abstract and references. The first section, which included the literature review, took more than three pages.

List of Works Consulted

A list of works consulted cites only those books, articles, papers, presentations, scripts, reports, or other sources that you used to develop your document. You might consult 50 works to produce your own document, although your preliminary bibliography provided you with 2,000 works to choose from. The 50 you selected, or consulted, directly helped you

The purpose of the current study was to examine, in detail, situations in which women attempt to use influence to have sex with a reluctant male partner. This has been a relatively neglected area of research, perhaps because this situation is not part of the traditional sexual script for heterosexual dating relationships. The traditional sexual script prescribes specific, socially accepted behaviors for men and women (Gagnon, 1990; Mosher & Tomkins, 1988). According to this script, men are expected to be the initiators of sexual activity and to maintain a constant search for sexual opportunities. In disagreement situations in which the man desires a more intimate level of sexual activity than his dating partner, the man is expected to influence the woman to have sex by whatever means are necessary, even if the form of influence in coercion (Check & Malamuth, 1985; Korman & Leslie, 1982). In contrast, women are expected to demonstrate little desire to engage in sexual activity and to control the amount of sexual access a dating partner is permitted (Check & Malamuth, 1985; Clark & Hatfield, 1989; Korman & Leslie, 1982; McCormick, Brannigan, & LaPlante, 1984). Women are expect to influence men to avoid sex, not to have sex (Clark & Hatfield, 1989; McCormick, 1979). . . .

Although few researchers have directly examined women's use of influence with a reluctant partner, it is clear from past research that women at times attempt to influence men to engage in sexual activity. For example, Perper and Weis (1987) analyzed seduction essays written by 29 U.S. and 48 Canadian women. Six percent of the essays described the women doing something with the hope that the man would "catch on" despite his initial nonresponsiveness to the woman's seduction attempt. Further, approximately 27% of the essays described dealing with a "recalcitrant male" who needed encouragement. However, these scenarios depicted a woman's use of influence when initiating the first sexual encounter, not necessarily situations in which a partner has rejected her sexual advance. In another study, Muehlenhard and Cook (1988) found that 93.5% of the 507 male college students they surveyed reported having been pressured to engage in some form of sexual activity on at least one occasion. Unfortunately, it is impossible to determine what percentage of these incidents was a result of pressure from a woman and what percentage was a result of pressure from other sources, such as self, peers, or a male partner. It is also unclear whether the man had communicated his reluctance to his partner. . . .

A related question has to do with the conditions under which disagreements characterized by a woman desiring a more intimate level of sexual activity than does her male partner and the subsequent use of influence behaviors occur. O'Sullivan and Byers (1992) found that deviation from traditionally scripted behaviors was most likely to occur in steady dating relationships. This finding is consistent with the theory that women adopt alternative scripts when they have assessed the likelihood of rejection by their dating partner to be low (Roche, 1986). Therefore, we predicted that reports of compliance and the subsequent attempts to influence a reluctant

Figure 17.3 Excerpt from the Introduction and Literature Review in a scientific article.

partner to have sex would be more likely to characterize steady dating relationships than first dates or casual dating relationships. In addition, the relationship between reports of sexual disagreements and a number of dating and personality variables that have been found to be related to men's attempts to influence reluctant women to have sex was investigated (Byers, 1988; Byers & Eastman, 1979: Byers & Eno, 1992; Kanin & Parcell, 1977; Koss, 1985; Muehlenhard & Linton, 1987). Sexuality and sex-role attitudes, history of engaging in the disputed level of sexual activity with a partner, and whether the participant had engaged in any sexual activity with a partner prior to the disagreement were examined....

Reprinted with permission of the Society for the Scientific Study of Sexuality from *The Journal of Sex Research*, 30(3), August 1993, pp. 270-282.

Figure 17.3 (*Continued*)

develop your understanding of the subject matter and the type of information that had been previously published. A list of works consulted illustrates the variety of sources you used to develop your ideas, regardless if you used any material from the sources. In fact, when you refer to most style manuals, the list of consulted works indicates sources that you used to get ideas for your work, but that you didn't quote, paraphrase, or summarize in the document you've written. Usually, a list of works consulted is longer than a list of works cited, because you review many sources before selecting those that you'll actually use in your document or as the basis of your hypothesis.

List of Works Cited or References

The list of works cited represents only works from which specific ideas have been taken. If you quote, summarize, or paraphrase information from a work, you must cite it. If you reproduce someone's work, such as a photograph or a sample from a document, you must cite the person who created the work and/or the publication from which the work was taken. If you interview someone or take someone's expression of an idea from a network, or from any other source, you must cite it. In a formal document, such as a scholarly article, a reference book, and a formal report, this list often takes the place of a note page and is a formal bibliography.

Depending upon the style manual you use, the list of works cited might be called Works Cited or References, for example, or it might be given another, slightly different title. In any case, this list, which most

The owner of a small business faces many pressures to keep a business operating day after day. No matter how hard the owner works, the business stands a good chance of failure, just because it is a new business that has not yet been around long enough to become established in the community. According to the Bureau of Labor Statistics, "the number of companies with fewer than 100 employees increased about 8 percent" (1) from 1980 to 1984. The competition for new businesses is extremely high with such a growing number of new businesses each year, and the entrepreneur who starts a business must be prepared to deal with this competition or fail, as so many new businesses inevitably do.

Figure 17.4 An in-text citation style.

often appears at the end of the document, provides a complete citation for every work cited in the body of your document. The items are formatted in the appropriate style, but the items cited are arranged alphabetically. According to some style manuals, they may also be numbered sequentially.

Within the document you've written, you must have an in-text citation that corresponds to a complete source citation in the bibliography. Again, depending upon the style manual you use, you may be required to cite sources in one of several different ways. For example, some in-house and professional associations' style manuals may require you to note the source you've used by placing in parentheses the number of the formal citation listed in the bibliography (see Figure 17.4).

At the end of the document shown in Figure 17.4, the list of works cited includes a notation for the source listed in the report as (1) (see Figure 17.5). This format is required by some journals and conference committees. If this example were used in MLA style, the example would look like Figure 17.6. If the reference manual required APA style, the example might look like Figure 17.7.

No matter which format is used, the idea behind the list of works cited remains the same: Only those sources quoted, paraphrased, or summarized directly in your document are listed as works cited.

List of Works Cited

(1) Freedman, David M. "Tailored Services for Small Firms." *Nation's Business* 74 (April 1986): 22-24.

Figure 17.5 A Works Cited citation sample.

The owner of a small business faces many pressures to keep a business operating day after day. No matter how hard the owner works, the business stands a good chance of failure, just because it is a new business that has not yet been around long enough to become established in the community. According to the Bureau of Labor Statistics, "the number of companies with fewer than 100 employees increased about 8 percent" (Freedman 22) from 1980 to 1984. The competition for new businesses is extremely high with such a growing number of new businesses each year, and the entrepreneur who starts a business must be prepared to deal with this competition or fail, as so many new businesses inevitably do.

Works Cited

Freedman, David M. "Tailored Services for Small Firms." *Nation's Business* 74 (April 1986): 22-24.

Figure 17.6 An in-text citation and Works Cited in MLA style.

The owner of a small business faces many pressures to keep a business operating day after day. No matter how hard the owner works, the business stands a good chance of failure, just because it is a new business that has not yet been around long enough to become established in the community. According to the Bureau of Labor Statistics, "the number of companies with fewer than 100 employees increased about 8 percent" (Freedman, 1986, p. 22) from 1980 to 1984. The competition for new businesses is extremely high with such a growing number of new businesses each year, and the entrepreneur who starts a business must be prepared to deal with this competition or fail, as so many new businesses inevitably do.

References

Freedman, D. M. (1986, April). Tailored services for small firms. *Nation's Business, 74*, 22-24.

Figure 17.7 An in-text citation and References in APA style.

Notes

When you have a bibliography, there must be some kind of in-text citation for that source in the body of your paper; otherwise, your readers won't know which source provided which piece(s) of information. Most often, that in-text citation is in the form of a number or short citation, as shown in the previous section. But when you want a more descriptive comment in addition to a short citation, or you work with a style manual that requires you to use a footnote style, you will need to work with notes, either a footnote or an endnote.

A *footnote* may be necessary to provide an in-text citation, but it may also provide additional information that some readers may need. For example, a discursive footnote may include a history or background information some readers new to the subject may need; a footnote might also provide additional descriptions, references for further reading, a comment about the source, or a translation of foreign terms or phrases. When readers need to see the information at the bottom of the page, the noted material appears in a footnote.

Some reference manuals specify that noted information appear as *endnotes* following the section, chapter, or report. These notes are similar to footnotes, because they are numbered to correspond to a superscript number in the text, and they may include the same types of information, but they are placed on a separate page in numeric order at the end of the section, chapter, or report.

Most publishers, either within a company, on behalf of a professional association, or for an international, paying readership, now prefer the use of in-text citations and full bibliographies. Authors/researchers are encouraged to limit the number of footnotes or endnotes in their work, and the information that may have been deemed worthy of a special note is now incorporated into the text or relegated to an appendix.

Before you submit any document for publication, you should get a copy of the publisher's specifications for documents and use the note and bibliographic styles appropriate for that publication.

Summary

As a researcher in technical communication, you'll have plenty of practice working with bibliographies for your projects and reviewing others' bibliographies to point you toward sources you might use. The best advice we can give is this: Methodically keep track of the sources you use or plan to use. Then check with the appropriate style guide to provide you with the information you need to document your work for publication. And always document your work, not only to avoid plagiarism but to conduct your research ethically.

Research Activities

1. Get a copy of the specifications used by a publication you read frequently. Then become familiar with the style of writing in-text notes and bibliographic information.
2. Create the appropriate type of bibliography for a document you're currently working on. Double check your sources to ensure you've accurately recorded the bibliographic information.
3. Develop a preliminary bibliography of sources you might use for a future research project.

For Further Reading

The following sources are only a few of the many in-house, professional, and association style manuals and related guides that can help you learn more about specific bibliographic styles. You should consult several style manuals and publishers to determine which style manual(s) and publishers' guidelines apply to your research and publications.

American Society for Microbiology. 1991. *ASM Style Manual for Journals and Books.* Washington, DC: ASM.

American Psychological Association. 1994. *Publication Manual of the American Psychological Association.* 4th ed. Washington, DC: APA.

Council of Biology Editors. Committee on Form and Style. 1972. *CBE Style Manual.* 3rd ed. Washington, DC: CBE.

Kirkman, John. 1992. *Good Style: Writing for Science and Technology.* New York: E & FM Spon.

CHAPTER EIGHTEEN

Conducting Additional Research

Research is one of the most important unifying characteristics of the technical communication profession. Whether you are in manufacturing, technology, medicine, graphics, education, training, or another technical or scientific area; whether you write manuals, proposals, reports, books, brochures, newsletters, articles, or other documents; whether you present your work online, on paper, both, or primarily through conferences or workshops, the motif that connects you with all others in the field of technical communication is research.

As professionals, we create information to meet specific communication objectives. But seldom are the objectives we meet the result of research we've limited to a particular assignment. Usually, we meet those objectives by calling upon all the resources we have at our disposal: our history of work experience and education; technical and scientific sources of information; human, paper, and online sources of information.

Because those resources are numerous, to access them efficiently and use them properly requires that each of us engaged in the profession be skilled at each technique for information gathering. Another interesting and exciting characteristic of the profession is that this skill involves not only knowing how to find the information required to complete certain kinds of communication assignments; it also requires knowing how to select and test the media available for presenting information to the people who ultimately need or want it.

In this book, we have emphasized a number of points about research that we hope will provide you some new approaches that will help you become more efficient and productive as you conduct research. This chapter is designed to lead you both out of the book into a new era of research methodology, and to lead you back into the book, which we hope will become a reference tool, to begin at the beginning with a new research project. In this chapter, then, we note some reasons why research is so important, summarize many strategies for conducting research, and provide some suggestions about what you may want to consider doing with your research.

Rewards and Consequences: Why Research is Important

For many reasons, it would be nice (as well as restful and relatively easy) if we could look at our work, conclude that we are good writers, and weave our ways through our professional careers based on that writing skill. But it won't happen. Both the ideas we present and the methods we use to present them are in constant flux. Thus, we are in a profession that, at its dullest, is exciting, for the results of what we do will have an impact upon people.

Think about it. The surgeon might transplant a heart, but to prepare for that transplant he or she read books, articles, and papers; went to conferences; observed other doctors; and experimented on cadavers and/or with computer simulations. When the surgeon steps into the operating room for any operation, what he or she does is the result of all the research he or she has done to that moment.

The astronaut might pilot a spaceship through the universe, reach out and touch the moon with various scientific equipment, and return that ship and its crew to earth safely. When he or she does, it is because of the experience and knowledge of all the research that went before that mission.

On gift-giving holidays, a couple may put their children to bed early and pull out the pieces of a new, red wagon. They look at the list of tools they'll need, match the parts in the package against the parts list, sort the various pieces, look at the exploded diagram to ensure they have a vision of where the pieces will fit, and then begin to read the instructions, assembling each piece as they are told. They avoid plier pinches and hammer bruises. They succeed quickly, go to bed early, and sleep soundly, awakening to happy noises of children pulling that wagon through the house. Neither they—nor their children—will care that every part of those instructions was meticulously researched; crafted for content, organization, and style; and tested for usability and accuracy. Nevertheless, what they have is the result of research.

When the research in each of these sample cases has provided the people creating the literature with accurate content, appropriate ideas about organization and style, correct suggestions about the use of color and graphics, and the correct medium, everyone is happy. When the research or the application of that research is inaccurate or inappropriate, the results may range from minor annoyance to catastrophe.

Where Should You Begin?

Research is both general and specific, goal oriented and objective oriented, but it is always prompted. When you subscribe to a particular magazine, read a particular book, monitor a particular electronic newsgroup or mailing list, you don't always have a defined reason for that choice. You have a curiosity about the subjects covered, a personal or passing professional interest. But you are responding to a prompt—something in you wants you to do better or more, know more, accomplish more, or just be aware of more. Specific or not, you're accomplishing goals for yourself, and you are deeply engaged in the ongoing process of research.

On the job, you are probably more specifically objective oriented. Perhaps your job is to create the user manuals for new software, so your research is constantly aimed at meeting your objective better, faster, and more efficiently. Perhaps your objective is to create marketing literature for a new product or online information for an automobile repairperson, or to put all of your company's information online. Whatever the objective, what you do and how you decide to do it will be the result of a number of factors, some of which you can clearly define, others which will be less clearly defined.

Know the Sources an Assignment Requires

When you are given a communication assignment, you'll probably define clearly many questions, even if you can't readily provide answers, which will help you plan the assignment. Those essential questions about any communication assignment may include, for example, the following:

1. What is the product or service you're communicating about?
2. What do you already know about the product, service, or information you're providing?
3. To whom are you communicating? What will they do with the information? What environment will they be in as they use the information? What kinds of pressures will they be under as they use the information?
4. Given what you believe you know and what you believe your audience needs to know, what do you need to find out about the content?
5. Where can you discover the information you need to know about the content? What sources are readily available? Are there are other sources? How do you find out if there are other sources?
6. Given your audience and the content you know, what is the best medium or are the best media for presenting that content? What do you know about using those media? Where can you find out more?
7. What are company policies, standards, procedures, and so on, that will affect the product you create, your time in creating it, and your procedures for both gathering information and creating the final project?

Plan the Project and the Research You'll Conduct

With the answers to these questions, you can plan, minimally, the following parts of your documentation process:

- Timeline
- Tasks: research, writing, reviews, production, testing/validation
- Restrictions (standards, policies, procedures, budget, personnel, equipment)
- Tentative content
- Tentative format, organization, and style

Placing all the information you know into an organized plan allows you to see exactly what you know. Comparing what you know to what you need to do allows you to begin to make preliminary notes about possible sources of information. That is, you have the beginnings of a preliminary bibliography, which includes sources to help you discover content, but which also can help you determine how to handle format, organization, style, reviews, testing/evaluation, and production of your final project.

Frequently, some sources of information may be required by company policy. You might be required to talk to product designers, for example, as you develop documentation about a new product. As frequently, however, the apparent sources are not the only sources available nor, for that matter, are they always the best. Thus, you always need to be aware of other possible sources, or at least know how to discover if there are other sources.

As you use sources or determine which sources to use, you'll constantly evaluate those sources. All sources don't provide the same kinds of information, nor the same quality of information. As you consider what information you need or where you may find available information, be ready to evaluate the information you'll find and its usefulness to you.

As the sources differ in quality, they also differ in methods of access. Certainly, constant sources of information are books, articles, and other kinds of printed information available from libraries. But those are minimally secondary sources for you. Sometimes the best information you can get is that which you discover through experimentation and observation. At other times, the best information is provided by SMEs, those people who are authorities about the information you need. You need to know who probably has the best information.

Knowing where the information is and being able to make some tentative judgments about the quality of the information you'll get from the various sources is only part of the process. To get the information, you need to know how to access the sources. Knowing how to read bibliographies is helpful for locating printed information. But you also need to write letters of inquiry soliciting information or the opportunity to solicit information from people. You need to plan, conduct, and follow up interviews.

Throughout the research process you need to keep accurate records of what you've done, what you've read, and with whom you talked. Your process should be meticulous, and the record of that process should reflect your care.

Writing a document is not the end of the research process, either. In fact, in many cases, writing may be more of a beginning to the real research challenge. Information has a purpose, an objective. Evaluating whether it has achieved that objective is a vital part of research. Through reviews, testing/evaluation of content, usability testing of the document's format, structure, and organization, the research process continues. Thus, as an effective researcher, you need to know how to organize a review process and conduct these tests, or at least monitor them, and evaluate the results.

All research for any project is not just a result of that focus. Many decisions you make about content and style are the results of general, ongoing research that has no specific focus on creating communication; instead, they are the result of a continuous effort you make to stay informed about what's going on in your field.

Suggestions for Staying Informed

Staying generally informed about what's going on in your field of work may, in many ways, be more difficult, more complicated, than conducting research to complete a specific project. At least, when you have an assignment, you clearly define your objectives for that assignment. Staying informed is not a clearly defined objective; instead, it is an elusive goal. It requires that you stay informed not only about what's currently happening, but that you are aware of possible changes. That is, not only do you have to keep up with sources that you know exist, but you also know how to discover when new sources are about to exist, and what each of them means to you. And staying informed means that you should actively participate in creating, instituting, and evaluating change.

Those are laudable goals, but how do you hope to achieve them? You can become involved in several professional activities to help you keep informed and to help focus the direction of your profession.

Participate in Professional Societies

Participating is more than just joining a professional society, although membership is the first step in participation. Professional societies like the Society for Technical Communication provide a number of opportunities for you to conduct both general and specific research. Through their journals, you can read theoretical and practical discussions about what's going on in the profession. If the society publishes books, you have a publisher available whom you know has the same vested interest in your field as you have. The publications are probably, therefore, focused on the kinds of information that you need. Through newsletters, you get current information about the profession, new media, new publications, new theories and applications, and announcements of courses, seminars, conferences, and publications that can be helpful in your various research projects.

Reading about theories and applications, however, is not always sufficient information. You should attend local, regional, national, and international conferences sponsored by the societies. Attendance at these conferences provides you with multiple opportunities to collect information. Not only do you hear about current theories and practices, but you also have the opportunity to ask questions and to speak both informally and formally with the speakers about their areas of interest. It's also an excellent opportunity to discuss your current and future research projects with those people you know have at least a related interest.

You should also present papers at conferences. You may think that presenting ideas is the result of research, not part of the research process, but it can be both. A paper is the presentation of ideas you have formulated to the point when the paper is presented. Although the papers should be focused and show meticulous and thorough research, you can also regard the presentation as an opportunity to learn more. Presenting the paper provides you the opportunity to learn the questions an audience has about your research, to see general reactions to your research, and to get comments for improving either your theory or application. Presenting a paper is an opportunity for testing and evaluating, for solidifying some ideas and changing others. The information you get is both fresh and focused, for it comes from people who share your interest in the subject.

Conferences also provide you the opportunity to respond to other speakers and thus gather information for your research. Meeting and talking with speakers and attendees also helps you build a network of subject matter experts, people you can contact when you need information, and people who will contact you for information.

You can also conduct both general and specific research by holding an office or managing committees. As an office holder in a society, you not only know what's going on in the society, but you also have the opportunity to influence the direction it's going. The same is true if you manage a committee. Officers and members of committees have an obligation not only to direct the profession today but to anticipate the future of the profession. Through your work in these committees and special groups, you have the opportunity to conduct both primary, current research and to plan for anticipated research.

Read the Literature

The literature in any field is composed of books, journals, articles, newsletters, brochures, and so on. Some literature is designed to provide bibliographic information; other literature provides theory or application. But the literature is more than just the content of a discussion. It also includes, for example, reviews of books and letters to the editor. Book reviews are certainly one source of information to help you decide whether you want to buy a book about a particular subject. But they provide you other information. On the simplest level, you discover what is being published, and therefore know of new books that are available and perhaps useful, regardless whether the reviewer liked the book. Reviewers discuss the contents of the particular books and frequently talk about approaches taken in the books, not just about the organization of the information but the content as well. These reviews can help you discover new theories people are working on or new approaches to old theories.

Surf the Net

Select one or more discussion groups or mailing lists that sound like they discuss topics that relate to your interests or occupation. Frequently, these sources will facilitate a discussion of issues important to your work, but they also provide other opportunities to ask questions, gather information, and just listen to what interests others. Numerous people will respond to questions you might have about specific subject matters. Sometimes the answers are content specific, and sometimes they simply provide other sources that can help you find answers to your questions. In either case, the groups are made up of people who at least share a related interest in your subject matter. Through these networks you can discuss with SMEs your various research issues. But, just as important, you can create a network of people of like mind or like interest whom you can contact individually and who individually can contact you to pursue various research projects.

Using the network is more than finding a few newsgroups and lurking about on them, or even participating in them. The network is still not clearly defined, either as what it really is or as what it really contains.

Thus, experiment with the net. Don't just read about gophers or the World Wide Web; push keys and travel. If, in fact, the network is the information superhighway, then it's your superhighway, and you not only have the right but the obligation to travel it. Travel and record the process and results of your travel. Not only will the travel itself provide you some general information about how you can use the net and about what's available on it, but it might also lead to some very specific sources of information for your current research.

Don't be concerned about accessing a database where you shouldn't be, or about someone monitoring your travels (unless of course you use your employer's network and that employer has indicated there are restrictions of use). Any place you end up is where you have a right to be, so travel, experiment, and gather the information you need.

Define the Literature

Reading the literature, participating in professional societies, and lurking on the Internet are all ways you can stay abreast of the information produced in your field. As you follow leads you may do so with the single aim of staying current with that information. However, as you read, you should also consider that if you contribute to the literature, you contribute to your own research, and you help define the literature.

Generally, when we think about what we read, we are reminded of some class long ago in which we learned that people publish original material. That is in many ways true, but it is important to understand all the word *original* means. *Original* certainly means new content, but it also means the following:

- Delineating new approaches and their results to existing information.
- Adding new or corrected information to existing content.
- Refuting existing information.
- Supporting existing information through additions of new processes or procedures.
- Designing new applications for existing information.
- Presenting existing information in new forms or formats, styles, or structures for different audiences.

As you participate in various research activities to keep yourself generally aware of what's going on in your field, always do so with a

1. List ideas you read that you might want to learn more about.
2. Note ideas you agree or disagree with.
3. Evaluate the ideas for accuracy, thoroughness, timeliness, completeness.
4. Evaluate the medium for its effectiveness either in presenting the ideas appropriately or in reaching the various audiences who might be interested in the idea.
5. Compare the ideas with your own knowledge and experience. Note areas where you agree, and emphasize areas where you may not agree.
6. Formulate your general ideas for further research.
7. Evaluate the mental or written notes you've made about what you've read.

Figure 18.1 A critical analysis of what you read.

critical mind. That is, look at the information as part of a foundation for a process that never ends. Always consider how you might contribute to that process (see Figure 18.1). Your general process of considering how you'll contribute to the literature should include the points detailed in Figure 18.2.

What else can you do with your ideas? We began the previous section by discussing ways you can evaluate the information you read to see how you might contribute to the literature. Thus, it is always in your best interest to know who is potentially interested in your ideas and how you might present those ideas.

Contacting Publishers

Through your various readings, you should develop a familiarity with journals and books that appear to focus on the kinds of information in which you are primarily interested. As you read, you should also be evaluating them as sources of information. When you have an idea you believe you want to pursue, your evaluation of possible publication sources becomes a focus of your research.

Limit the sources you consider initially by evaluating your own information. Do you need to test it by presenting it orally to an audience?

- Evaluate the sources as you use them.
- If you discover a possible weakness in methodology or conclusions, then formulate a theory of your own to pursue.
- Brainstorm, freewrite, discuss, or cluster your ideas.
- Organize your results: sort and categorize them.
- Reevaluate your theory in terms of what you know.
- Consider an audience for your ideas: Who are they—academics, writers, artists, others? Why would they care to know about your ideas—for their own research, application, replication, general personal, or professional interest? For each audience, what can you assume they know? What can you assume they don't know?
- Instead of trying to publish the ideas you've formulated, what are the possible ways you can apply them—on the job, personally, as a member of a community or organization?

Figure 18.2 Your contribution to the literature.

Is timeliness an issue? If your answers are yes, then you need to consider presenting your information at a conference or a meeting sponsored by the professional society dedicated to the profession that your information affects.

You can contact a professional society for conference or meeting information. Read Calls for Papers or Announcements of Meetings. Request the material that details requirements for submitting a proposal to present an abstract or a paper.

If your information needs a larger audience, is too limited in scope to be enough for a book, or is by content or approach more appropriate for a written medium, consider a journal or magazine. Review the journals and magazines you read to discover the kinds of information they publish, whether they have thematic issues dealing with certain subject matters. Someplace in each journal will be either information that details the guidelines for publishing in the journal or that tells potential authors how to get those guidelines. Read those sections carefully, and if necessary, write a letter of inquiry soliciting the guidelines you need.

If you believe your information is appropriate for a book, review any similar books currently on the market. See who publishes them. Read reviews about them. Read the prefatory information to determine who the books' audiences are, what the books' stated purposes are, and any other information that tells you about the books' markets.

Evaluate possible publishers, and plan your contact with them. Your first contact may be just a short letter of inquiry in which you briefly indicate who you are, your book idea, and your reason for contacting that particular publisher. Once you have heard from the publisher, you will probably then be required to submit a more detailed prospectus.

Research Requires Time And Money

On the job, the research you do will be limited by the requirements of the project; your company's imposed limitations of time, personnel, budget, standards, policies, and procedures; and your other work requirements. The same kinds of restrictions exist when you are completing research about ideas that you wish to pursue outside work or as a related part of work for which there is no company support. Thus, it is also an important part of your research to know not only who might be interested in the ideas you want to pursue, but also who might fund those ideas.

Sources of funding are numerous and varied, and for the most part potential funders don't do a lot of direct advertising. As a continuing part of your research, therefore, you should stay aware of possible funding sources.

Frequently, professional societies offer funding for people to conduct research that has the potential to improve the profession as a whole. The Society for Technical Communication, for example, makes grants available for research into new technologies, management strategies, ways to improve the recognition as well as the practice of technical communication, and ways to provide individuals and chapters opportunities to present conferences and meetings, or to pursue other activities designed to benefit those chapters locally.

Companies and other regional organizations fund projects that improve their communities. The programs range from community arts endeavors to those designed to aid the elderly or the underprivileged.

City and state governments also fund projects that are designed to improve local communities and thus the state. The federal government offers multiple programs that are designed for the sole purpose of providing sources of funding for individuals, groups, and companies to pursue innovative research and the subsequent presentation of that research.

Private research institutions also support research into the areas of study with which they are concerned. Certainly, most of us are familiar with the research institutes designed to provide information about government programs, but those are not the only institutes, nor are they the only sources of funding. Getting funding for your research is a two-step, continuous process:

1. Know where funds are available.
2. Know how to write the grant proposals.

One way to know where grantors are located and which grants are available is always to read the literature devoted to your profession, not only for its content, but also to see if it is sponsored. If an article or a book, for example, is the result of sponsored research, the sponsor will be noted clearly, usually in the prefatory or introductory parts of the material.

Also read the literature for announcements of monies available for research. Again, the Society for Technical Communication announces the availability of grant monies in their periodic newsletters, in their journal, and through minutes of the international board meeting. Journals and magazines frequently have sections devoted to listing events like forthcoming conferences, meetings, and publications, but they also list sources of funding.

City and local governments, as well as libraries, have access to and information about grants that affect them. Contacting governmental agencies can at least provide you ideas about others you should contact.

The Internet has become yet another source of information regarding funding sources and, in some cases, requirements for being funded through those sources. The number of agencies and the information about those agencies that is available on the Internet continues to grow. Thus, your best approach now is to access a gopher and complete some keyword searches to see what appears. You might also ask those people who participate in various mailing lists or newsgroups you read for access sites.

Along with, and as a foundation for all other searches, you should still consult the hardcopy literature to see which grants are available. The *Federal Register* lists, for example, grants that the federal government sponsors. In addition, many agencies are devoted to funding innovation in research and application conducted by small businesses. Contacting your local Small Business Association is an important first step in locating these sources.

Different granting agencies have different requirements for the way they want grant proposals created. For some, you may need to do no more than write a letter detailing your ideas. Other grantors, though, require detailed proposals, whereas still others require a combination of a detailed proposal, plus the completion of multiple forms. Some grantors may request that you submit proposals online, whereas others want only paper-printed proposals. As you consider the possibility of writing a grant to get your research funded, consider all the possible sources and learn how to submit the grant proposals to them.

Summary

Conducting research is not only exciting, it's required. Technical communicators have a daily responsibility to conduct research appropriate to their jobs, and they have a further obligation as members of a dynamic profession. Research is necessary to continue to expand our definitions of what technical communication is and should be, and how it can be made better. To be an effective researcher, you need to become or remain actively involved in the profession and to seek avenues of presentation, publication, and future support for your ongoing research.

Research Activities

1. If you are not currently a member of a professional society, conduct research into appropriate associations and societies that may provide you with opportunities to conduct research and to share your work with others. If you are a member of a professional society, investigate ways that you can more actively participate in research activities associated with that society.
2. Conduct an Internet search of potential grantors for the type(s) of research you do. Who are potential grantors? When do they offer grants? What are their requirements? You might begin a file, in hard- or softcopy or both, about grants and grantors.
3. Set professional objectives for the next year. Where can you attend conferences? Where can you present papers? Which journals should you thoroughly read? How can you use computer networks and databases to best advantage? Develop a plan that you can follow to increase the number of your research activities.

APPENDIX A

Master Bibliography of Sources Listed in This Book

Aboba, Bernard. 1993. *The Online User's Encyclopedia: Bulletin Boards and Beyond.* Reading, MA: Addison-Wesley.

Achterberg, Cheryl. 1989. "Tips for Writing Abstracts." *Journal of Nutrition Education* 21(5) (September): 197.

Aiken, Milam, Jay Krosp, and Ashraf Shirani. 1993. "Electronic Brainstorming in Small and Large Groups." *Information Management* (September): 141-48.

Allen, Nancy. 1993. "Learning E-mail Skills." *Technical Communication* (November): 766-68.

Alspach, Ted. 1995. *Internet E-mail Quick Tour.* Chapel Hill, NC: Ventana Press.

American National Standards Institute. 1979. *American National Standard for Writing Abstracts. Secretariat: Council of National Library and Information Associations.* (Standard Z39.1401979, a revision of Z39.14-1971.) New York: The Institute.

American Psychological Association. 1992. "Ethical Principles of Psychologists and Code of Conduct." *American Psychologist* (December) 47: 1597-611.

American Psychological Association. 1994. *Publication Manual of the American Psychological Association.* 4th ed. Washington, DC: American Psychological Association.

American Society for Microbiology. 1991. *ASM Style Manual for Journals and Books.* Washington, DC: ASM.

Appleton, Elaine L. 1993. "Put Usability to the Test." *Datamation* (July 15): 61-62.

Babbie, Earl. 1995. *The Practice of Social Research.* 7th ed. Belmont, CA: Wadsworth.

Barker, Thomas T. 1989. "Word Processors and Invention in Technical Writing." *The Technical Writing Teacher* (Spring): 126-35.

Bell, Judith. 1993. *Doing Your Research Project: A Guide for First-time Researchers in Education and Social Science.* 2nd ed. Philadelphia: Open University Press.

Berners-Lee, Tim. 1994. "The World-Wide Web." *Communications of the ACM* (August): 76-82.

Boiarsky, Carolyn. 1992. "A Teaching Tip. Using Usability Testing to Teach Reader Response." *Technical Communication* (February): 100-02.

Borg, Walter R., Joyce P. Gall, and Meredith D. Gall. 1993. *Applying Educational Research: A Practical Guide.* 3rd ed. New York: Longman.

Branwyn, Gareth. 1994. *Mosaic Quick Tour for Windows: Accessing and Navigating the Internet's World Wide Web.* Chapel Hill, NC: Ventana Press.

Braun, Eric. 1994. *The Internet Directory.* New York: Fawcett Columbine.

Clandinin, D. Jean, and F. Michael Connelly. 1994. "26 Personal Experience Methods." Norman K. Denzin and Yvonna S. Lincoln, eds. *Handbook of Qualitative Research.* Thousand Oaks, CA: Sage.

Council of Biology Editors. Committee on Form and Style. 1972. *CBE Style Manual.* 3rd ed. Washington, DC: CBE.

Cragg, Paul B. 1991. "Designing and Using Mail Questionnaires." N. Craig Smith and Paul Dainty, eds. *The Management Research Handbook.* New York: Routledge.

Cremmins, Edward T. 1982. *The Art of Abstracting.* Philadelphia: ISI Press.

Curtis, Donnelyn, and Stephan A. Bernhardt. 1991. "Keywords, Titles, Abstracts, and Online Searches: Implications for Technical Writing." *Technical Writing Teacher* (Spring): 142-61.

"Databases: Some Options for End-user Searching in the Health Sciences." 1989. *Reverence Quarterly* (Spring): 395-406.

December, John. 1994. *The World Wide Web Unleashed.* Indianapolis: Sams Publishing.

Denzin, Norman K., and Yvonna S. Lincoln. (Eds.). 1994. *Handbook of Qualitative Research.* Thousand Oaks, CA: Sage.

Desai, Pranav Kiritjumar. 1994. *Brainstorming as a Concept Generation Methodology in Mechanical Design.* Master's thesis. Case Western Reserve University.

Di Vittorio, Martha Montes. 1994. "Evaluating Sources of U.S. Company Data." *Database* (August): 39-44.

Dragga, Sam. 1991. "Classifications of Correspondence: Complexity versus Simplicity." *Technical Writing Teacher* 18(1) (Winter): 1-13.

Drummond, Rik. 1994. *LAN Times E-mail Resource Guide.* Berkeley, CA: Osborne McGraw-Hill.

Dumas, Joseph S., and Janice C. Redish. 1993. *A Practical Guide to Usability Testing.* Norwood, NJ: Ablex.

Eager, Bill. 1994. *Using the World Wide Web.* Indianapolis: Que.

Ehrlich, Kate. 1994. "Getting the Whole Team into Usability Testing." *IEEE Software* (January): 89-91.

Elbow, Peter. 1992. "Freewriting and the Problem of Wheat and Tares." *Writing and Publishing for Academic Authors,* pp. 33-47. Joseph M. Moxley, ed. Lanham, MD: University Press of America.

Erwin, Edward, Sidney Gendin, and Lowell Kleiman. 1994. *Ethical Issues in Scientific Research: An Anthology.* New York: Garland.

Every Student's Guide to the Internet. 1995. San Francisco: McGraw-Hill.

Frederick, William C., and Lee Preston. (Eds.). 1990. *Business Ethics: Research Issues and Empirical Studies.* Greenwich, CT: JAI Press.

Gardner, Sylvia A. 1992. "Spelling Errors in Online Databases: What the Technical Communicator Should Know." *Technical Communication* (February): 50-3.

Goss, Larry D. 1987. "Techniques for Generating Objects in a Three-Dimensional CAD System." *Engineering Design Graphics Journal* (Fall): 29-35.

Green, W. T., L. V. Sadler, and E. W. Sadler. 1985. "Diagrammatic Writing Using Word Processing: Computer-Assisted Composition for the Development of Writing Skills." *Computing Teacher* (April): 62-4.

Grice, Roger A. 1993. "Usability and Hypermedia: Toward a Set of Usability Criteria and Measures." *Technical Communication* (August): 429-37.

___. 1994. "Focus on Usability." *Technical Communication* (August): 521-22.

Hackos, JoAnn T. 1993. *Managing Your Documentation Projects.* New York: John Wiley & Sons, Inc.

Hahn, Harley, and Rick Stout. 1994. *The Internet Complete Reference.* Berkeley, CA: Osborne McGraw-Hill.

Hall, Dean G., and Bonnie A. Nelson. 1987. "Initiating Students into Professionalism: Teaching the Letter of Inquiry." *Technical Writing Teacher* 14(1) (Winter): 86-9.

___. 1988. "The Letter of Inquiry: A Neglected Tool for Technical Classes." *Engineering Education* 78(7) (April): 695-99.

Halpern, Jeanne W. 1988. "Getting in Deep: Using Qualitative Research in Business and Technical Communication." *Journal of Business and Technical Communication* (September): 22-43.

Hayes, Brian. 1994. "The World Wide Web." *American Scientist* (September): 416-20.

Hernon, Peter. 1989. "Government Information: A Field in Need of Research and Analytical Study." *United States Government Information Policies,* pp. 10-13. Charles R. McClure, Peter Hernon, and Harold C. Relyea, eds. Norwood, NJ: Ablex.

Hogan, J. B. 1983. "Statistical Doublespeak: The Deceptive Language of Numbers." *The Technical Writing Teacher* (Winter/Spring): 126-29.

Jobber, David. 1991. "Choosing a Survey Method in Management Research." *The Management Research Handbook.* N. Craig Smith and Paul Dainty, eds. New York: Routledge.

John, Nancy. 1994. *The Internet Troubleshooter: Help for the Logged On and Lost.* Chicago: American Library Association.

Jonassen, David H., and Heinz Mandl. (Eds.) 1990. *Designing Hypermedia for Learning.* Series F: Computer and Systems Sciences, Vol. 67. Berlin: Springer-Verlag.

Kirkman, John. 1992. *Good Style: Writing for Science and Technology.* New York: E & FM Spon.

Krauhs, Jane M. 1993. "Extend Your Confidence Limits in Writing about Statistics." *Technical Communication* (November): 742-43.

LaPlante, Alice. 1993. "'90s Style Brainstorming." *Forbes* (October 25): 44-61.

LeCompte, Margaret Diane, Judith Preissle, and Renata Tesch. 1993. *Ethnography and Qualitative Design in Educational Research.* 2nd ed. San Diego: Academic Press.

Lemay, Laura. 1995. *Teach Yourself Web Publishing with HTML in a Week.* Indianapolis: Sams Publishing.

Levine, Robert J. 1988. *Ethics and Regulation of Clinical Research.* 2nd ed. New Haven, CT: Yale UP.

Losee, Robert M., Jr., and Karen A. Worley. 1993. *Research Methods 101: Research and Evaluation for Informaton Professionals.* San Diego: Academic Press.

Lovell, Ron. 1988. "How to Write Textbooks for Fun, If Not for Profit." *Journalism Educator* (Fall): 34-5.

MacNealy, Mary Sue. 1990. "Moving Toward Maturity: Research in Technical Communication." *IEEE Transactions on Professional Communication* 33: 197-204.

___. 1992. "Research in Technical Communication: A View of the Past and a Challenge for the Future." *Technical Communication* 39(4): 533-51.

Mason, Nondita. 1994. *Writers' Roles: Enactments of the Process.* Fort Worth, TX: Harcourt Brace College Publishers.

Mauer, Mary E. 1994. "Brainstorming." *Writer* (December): 22-3.

McArthur, Douglas C. 1994. "World Wide Web & HTML." *Dr. Dobb's Journal* (December): 18-26.

McDowell, Earl E. 1991. *Interviewing Practices for Technical Writers.* Amityville, NY: Baywood.

McGuigan, Frank J. 1993. *Experimental Psychology: Methods of Research.* 6th ed. Englewood Cliffs, NJ: Prentice Hall.

McKnight, Cliff, Andrew Dillon, and John Richardson. 1991. *Hypertext in Context.* Cambridge: Cambridge UP.

Moxley, Joseph M. (Ed.). 1992. *Writing and Publishing for Academic Authors.* Lanham, MD: University Press of America.

Moxley, Joseph Michael. 1992. *Publish, Don't Perish: The Scholar's Guide to Academic Writing and Publishing.* Westport, CT: Greenwood Press.

Nadis, Steve. 1994. "Brainstorming Software: Can Your Computer Be Equipped with Artificial Creativity?" *Omni* (December): 28.

Newell, Rosemarie. 1993. "Questionnaires." *Researching Social Life*, pp. 94-115. Nigel Gilbert, ed. Newbury Park, NJ: Sage.

Nielsen, Jakob. 1993. *Usability Engineering.* Boston: Academic Press.

Notess, Greg P. 1994. "Lynx to the World-Wide Web." *Online* (July): 78-82.

Parker, Rachel. 1994. "Don't Test for Usability: Design for It." *InfoWorld* (July 18): 58.

Pinelli, Thomas E., and Rebecca O. Barclay. 1992. "Research in Technical Communication: Perspectives and Thoughts on the Process." *Technical Communication* 39(4): 526-32.

Plumb, Carolyn. 1992. "Survey Research in Technical Communication: Designing and Administering Questionnaires." *Technical Communication* (November): 625-38.

Queipo, Larry. 1991. "Taking the Mysticism out of Usability Test Objectives." *Technical Communication* (April): 185-89.

Reilly, Norman B. 1993. *Systems Engineering for Engineers and Managers.* New York: Van Nostrand Reinhold.

Richards, Thomas C., and Jeannette Fukuzawa. 1989. "A Checklist for Evaluation of Courseware Authoring Systems." *Educational Technology* (October): 24-9.

Robbins, R. 1987. "Helping to Make Reports Real: A Brainstorming Aid for Assignments in Technical Communication." *The Technical Writing Teacher* (Spring): 99-102.

Robson, Colin. 1993. *Real World Research: A Resource for Social Scientists and Practitioner-Researchers.* Cambridge, MA: Blackwell.

Rubin, Jeffrey. 1994. *Handbook of Usability Testing: How to Plan, Design, and Conduct Effective Tests.* New York: John Wiley & Sons, Inc.

Schatz, Bruce R. 1994. "NCSA Mosaic and the World Wide Web: Global Hypermedia Protocols for the Internet." *Science* (August 12): 895-901.

Schiltz, Michael E. (Ed.). 1992. *Ethics and Standards in Institutional Research.* San Francisco: Jossey-Bass.

Schultz, Susan I. 1993. *The Digital Technical Documentation Handbook.* Burlington, MA: Digital Press.

Schulzrinne, Henning. 1992. *Voice Communication Across the Internet: A Network Voice Terminal.* Amherst: University of Massachusetts at Amherst, Department of Computer and Information Science.

Schwarzwalder, Robert. 1994. "Searching at the Fringes: Finding Technical Information in Nontechnical Databases." *Database* (April): 100-04.

Shrader-Frechette, Kristin. 1994. *Research Ethics.* Lanham, MD: Rowman & Littlefield.

Sieber, Joan E. 1992. *Planning Ethically Responsible Research: A Guide for Students and Internal Review Boards.* Newbury Park, CA: Sage.

Simpson, Mark. 1991. "The Practice of Collaboration in Usability Test Design." *Technical Communication* (November): 526-31.

Skelton, T. M. 1992. "Testing the Usability of Usability Testing." *Technical Communication* (August): 343-59.

Slavens, Thomas P. 1985. *Reference Interviews, Questions, and Materials.* 2nd ed. Metuchen, NJ: Scarecrow Press.

Smith, Frank. 1988. "The Importance of Research in Technical Communication." *Technical Communication* 35(2): 4-5.

Spring, Marietta. 1988. "Writing a Questionnaire Report." *Bulletin of the Association for Business Communication* (September): 18-19.

Spyridakis, Jan H. 1991. "The Technical Communicator's Guide to Understanding Statistics and Research Design." *Technical Communication* (November): 207-19.

Stacks, Don W., and John E. Hocking. 1992. *Essentials of Communication Research.* New York: HarperCollins.

Stephens, Irving E. 1991. "Citation Indexes Improve Bibliography in Technical Communication." *Journal of Technical Writing and Communication* 117-25.

Sugg, David. 1993. "Putting Meat on the Bones—Cause-and-Effect Diagrams." *Plating and Surface Finishing* (November): 54-5.

Tipton, Martha. 1993. "Usability Testing for Multimedia." *CD-ROM Professional* (July): 123-25.

Tittel, Ed. 1994. *E-mail Essentials.* Boston: AP Professional.

Turlington, Shannon R. 1995. *Walking the World Wide Web: A Listings Guide for Multimedia Resources on the Internet.* Chapel Hill, NC: Ventana Press.

Varhol, Peter. 1995. *E-mail: Achieving Local and Global Communications.* Charleston, SC: Computer Technology Research Corp.

Vetter, Ronald J. 1994. "Mosaic and the World-Wide Web." *Computer* (October): 49-57.

Winkler, Earl R., and Jerrold R. Coombs. (Eds.). 1993. *Applied Ethics: A Reader.* Cambridge, MA: Blackwell.

Zikmund, William G. 1991. *Business Research Methods.* 3rd ed. Chicago: Dryden Press.

APPENDIX B

Some Mailing Lists for Technical Communicators

List Server Address	*Topic/Title*	*Subscription Command*
listserv@vm1.ucc.okstate.edu	technical writing	sub techwr-l fn ln
listserv@ncsu.edu	technical communication	sub TechCom fn ln
listserv@ubvm.cc.buffalo.edu	business ethics	sub buseth-l
listserv@austin.onu.edu	artificial intelligence and law	sub ail-l
listserv@knex.via.mind.org	CD-ROM	sub cdpub
listserv@vmd.cso.uiuc.edu	virtual reality	sub virtu-l
listserv@yalevm.ycc.yale.edu	desktop publishing	sub dtp-l
listserv@pltumk11.bitnet	science and education	sub appl-l
listserv@utkvm1.bitnet	software engineering ethics	sub ethcse-l
listserv@uga.cc.uga.edu	computer ethics	sub ethics-l
listserv@trearn.bitnet	image processing	sub image-l
listserv@utfsm.bitnet	latest computing advances	sub adv-info
listserv@wmvm1.bitnet	network law and policies	sub cyberlaw
listserv@uxa.ecn.bgu.edu	video production/operations	sub vidpro-l
listserv@umdd.bitnet	multimedia	sub imamedia
listserv@ulkyvm.louisville.edu	computer-assisted research	sub carr-l
listserv@psuvm.psu.edu	engineering problem solving	sub cre8tv-l
listserv@ubvm.cc.buffalo.edu	navigating the Internet	sub navigate
listserv@is.internic.net	network issues and events	sub net-happenings

listserv@ubvm.cc.buffalo.edu	network trainers list	sub nettrain
listserv@bitnic.educom.edu	information technology	sub edupage
listserv@mcgill1.bitnet	consultants	sub cons-l
listserv@rpitsvm.bitnet	research methodologies	sub methods

This list was compiled and shared on the technical communication mailing list in 1994 by Brad Mehlenbacher, North Carolina State University.

GLOSSARY

attribute	a value, characteristic, or quality that describes someone or something (such as male, Asian, Democrat, technical communicator)
brainstorming	a method of idea generation in which the researcher jots down as many ideas as possible within a limited time
browser	a program that lets researchers read information available in the World Wide Web and follow the hypermedia links they select
CD-ROM	computer disk-read only memory; a storage device that may provide interactive multimedia or hypermedia information
central tendency	in statistics, the typical characteristics of the data set
closed question	a type of questionnaire question in which the researcher lists the possible answers and thus limits respondents' choices of answers
computerized output microform	COM; an online catalog to find a library's microform collections, often stored on microfiche, including telephone directories, college and university catalogs, company reports, newspapers, and government documents
control variable	a constant variable against which other variables are compared
database	a collection of individual pieces of information that are accessible in many usable ways
dependent variable	a variable that depends on or is varied by an independent variable
descriptive statistics	statistics that organize and summarize data

dichotomous question	a question that can be answered with one of two possible answers (such as yes or no)
documentation plan	doc plan; a document that describes the research and production plan for the current project
endnote	a note similar to a footnote, often numbered to correspond to a superscript number in the text and placed on a separate page in numeric order at the end of the section, chapter, or report
essay question	a question that allows respondents to answer in their own words with as much detail as they want to provide
ethics	a code of behavior and values by which professionals abide
fill-in-the-blank question	a question that requires respondents to supply a word, number, phrase, and so on, in response
footnote	an in-text citation or a note at the bottom of a page that may include a history or background information, additional descriptions, references for further reading, a comment about the source, or a translation of foreign terms or phrases
freewriting	a method of idea generation in which the researcher keeps writing, without stopping, any idea that comes to mind
gopher	an online system to help researchers work with databases by providing them with a menu of options and then carrying out their research request
home page	an initial descriptive page about the contents of the networked materials placed on the World Wide Web
hypermedia	information presented in multiple media formats but allowing users to move from place to place within the information through hypertext links
hypothesis	a statement indicating the way things are expected to work, based logically on a theory
independent variable	a given variable (such as age)
inferential statistics	statistics that evaluate what you've learned about a group (the sample) and then generalize what you've learned to the larger population from which you took the sample
informational interview	a primary source of information; an interview conducted to learn more about a company or a subject

informative abstract	a detailed summary of a document's content
interactive software	a program perhaps created through authoring languages or a purchased application that allows users to respond to prompts and get feedback on their ideas; a program used in idea generation
letter of inquiry	a primary source of information; a formal letter written to a company or a person to solicit answers to questions, materials, or documents
local area network	LAN; a network of two or more computers connected within a small geographic area, such as within a company
mean	in statistics, the average score in the data set
median	in statistics, the point above which half the scores fall and below which half the scores fall
mode	in statistics, the most frequent score or value within the data set
multimedia	information presented in multiple media formats
multiple choice question	a question that can be answered with one of several possible answers provided by the researcher
multiple-user dimensions	MUDs; an online location where people can "discuss" topics, have informal conversations, and meet new people or characters; a virtual environment
open-ended question	a type of questionnaire question in which the researcher does not list possible answers and thus lets respondents provide answers in their own words
outlining	a method of idea generation in which the researcher outlines whatever ideas come to mind
personal experience	that which one has done before
personal knowledge	that which one has learned either from experience or other research
preliminary bibliography	a list of documents the researcher is consulting or plans to consult during the early stages of a long-term project; or a bibliography prepared for others to use in their research
primary source	information developed by the researchers themselves
proposal letter	another name for a query letter, which is sent to a prospective publisher

prospectus	a document similar to a query letter, but including more information about the proposed publication (such as a market survey that compares the proposed publication to similar publications already on the market)
qualitative research	research involving descriptive methods and the categorization of information
quantitative research	research involving methods that produce countable results
query letter	a letter sent to a prospective publisher to inquire about the possibility of publishing information about the current research
questionnaire	a primary source of information; a survey of questions given to respondents
range	in statistics, the difference between the highest score and the lowest score, or the difference between the highest score and the lowest score, plus 1
ranking question	a question that requires respondents to rank numerically their responses to provide a hierarchy of responses
rating question	a question that requires respondents to provide their responses regarding the quality of a product or service, the frequency of an occurrence or a behavior, or the degree of agreement with a statement
readability	the test for clear, precise, correct, manageable chunks of information (such as tests for grammatically correct language, the effective use of color in graphics, the selection of a large enough font for easy reading, the layout of pages so information is appropriately highlighted, the technically correct shades of meaning in word choice)
reference database	a database consisting of bibliographic information
secondary source	information created, printed, and/or published by someone else
source database	a database consisting of original, complete sources, usually documents
standard deviation	in statistics, the most common measure of variability; the average distance (the deviation) from the mean score
statistic	a fact that explains or describes the results of observation
statistics	the methods of making sense out of research data

subject matter expert	SME; an expert researchers consult about his or her technical or scientific area of expertise
tertiary source	information someone else created based on someone else's information
topical abstract	also called a descriptive or an indicative abstract; a brief description of a document that tells potential readers what kind of manuscript it is, what kind of information it's designed to provide, and how the information is organized
usability	the accuracy, readability, and functionality of information
validation	the test for accuracy; a verification of information
variable	a grouping of attributes (such as gender is the variable for attributes of male and female)
virtual reality	a created environment that mimics reality; a form of experiential information that can be used in idea generation
wide area network	WAN; a network of two or more computers connected within a wide geographic area, such as within a country
World Wide Web	WWW; information available through the Internet in multimedia formats

Index

Q

R

S

T

U

V

W